Curriculum, Environment, and the Work of C. A. Bowers

This edited volume extends ecological approaches to curriculum theory by recognizing and building on the contributions of the late Chet A. Bowers to curriculum and ecological studies globally.

Chapters provide in-depth explanation of Bowers' central contributions to the field, including his identification of the linguistic roots of ecological degradation; the need for school curricula to support sustainability; and the principles of cultural commons, eco-justice, and ecological intelligence. Building on these ideas and emphasizing the links between curriculum studies, social justice, and environmental education, the text illustrates how Bowers' ideas must now inform future approaches to schooling, teacher education, research, and Indigenous communities to guard against the global ecological crises we now face.

This text will benefit researchers, academics, and educators with an interest in curriculum studies, sustainability education, and environmental studies in particular. Those interested in the sociology of education, educational change, and school reform will also benefit from the book.

Audrey M. Dentith is Professor of Leadership Studies and Adult Education and Director of the Center for Teaching Excellence at North Carolina A&T State University, USA.

David Flinders is Professor of Education at Indiana University, USA.

John Lupinacci is Associate Professor of Cultural Studies and Social Thought in Education at Washington State University, USA.

Jennifer S. Thom is Associate Professor of Curriculum Studies and Mathematics Education at the University of Victoria, Canada.

Curriculum, Environment, and the Work of C. A. Bowers
Ecological and Cultural Perspectives

**Edited by Audrey M. Dentith,
David Flinders, John Lupinacci, and
Jennifer S. Thom**

Routledge
Taylor & Francis Group

NEW YORK AND LONDON

First published 2022

by Routledge
605 Third Avenue, New York, NY 10158

and by Routledge
2 Park Square, Milton Park, Abingdon, Oxon, OX14 4RN

Routledge is an imprint of the Taylor & Francis Group, an informa business

Library of Congress Cataloging-in-Publication Data
Names: Dentith, Audrey M., editor. | Flinders, David J.,
1955- editor. | Lupinacci, John, editor. | Thom, Jennifer S.
(Jennifer Susan), editor.
Title: Curriculum, environment, and the work of C.A. Bowers :
ecological and cultural perspectives / edited by Audrey M.
Dentith, David Flinders, John Lupinacci, and Jennifer S. Thom.
Description: New York, NY : Routledge, 2021. | Series: Studies in
curriculum theory | Includes bibliographical references and index.
Identifiers: LCCN 2021003932 | ISBN 9780367417864 (hardback) |
ISBN 9781032035628 (paperback) | ISBN 9780367822460 (ebook)
Subjects: LCSH: Environmental education. | Human ecology–
Study and teaching. | Social justice–Study and teaching. |
Education–Curricula. | Bowers, C. A.
Classification: LCC GE70 .C87 2021 | DDC 333.7071–dc23
LC record available at https://lccn.loc.gov/2021003932

ISBN: 978-0-367-41786-4 (hbk)
ISBN: 978-1-032-03562-8 (pbk)
ISBN: 978-0-367-82246-0 (ebk)

Typeset in Baskerville
by KnowledgeWorks Global Ltd.

This book is dedicated to Chet Bowers, extraordinary scholar, educator, mentor, and friend; and to Mary Bowers, a deeply committed teacher and friend who welcomed us into her classroom and home.

Contents

PART II

Curriculum of the Commons   89

Excerpt from: Bowers, C. A. (2016). *Reforming higher
education in an era of ecological crisis and growing digital insecurity.*
Anoka, MN: Process Century Press.

PART III

Ecojustice Curriculum   141

Excerpt for Part III from: Bowers, C. A. (2002). Toward
an eco-justice pedagogy. *Environmental Education Research,*
8(1), 21–34.

List of Figures

About the Contributors

Peter Cole is a member of the Douglas First Nation, one of the Stl'atl'imx communities in SW British Columbia, and also has Celtic heritage. He has taught at universities in Canada, the United States and Aotearoa, New Zealand, most recently as Associate Professor in Aboriginal and Northern Studies at the University College of the North where he was Chair of the Research Ethics Board. Cole has played key roles in the development of the Aboriginal & Northern Studies degree program at UCN; the Developmental Standard Teaching Certificate with four Vancouver Island First Nations communities to certify language teachers to teach their Indigenous languages in schools; and, while at Massey University in Aotearoa-New Zealand, was invited by Maori colleagues to participate in the reshaping of the pakeha Technology in the New Zealand Curriculum document into Hangarau: I roto i te Marautanga o Aotearoa, a curriculum based on Maori spiritualties, knowledges, and technologies. Beginning in January, 2001, Cole has been instrumental in initiating a dialogue with the Social Sciences & Humanities Research Council of Canada to be more inclusive and respectful of Aboriginal research protocols, epistemologies and methodologies.

Audrey M. Dentith is Professor in the Department of Leadership Studies and Adult Education and Director of the Center for Teaching Excellence (CTE) at the North Carolina A & T State University in Greensboro, NC. Her teaching and research interests include environmental studies in education, curriculum theory, feminisms and women's history in education. She is the recent editor (with Wendy Griswold) of *Ecojustice adult education: Theory and practice in the cultivation of the cultural commons* (2017). In recent years, she has organized institutes on ecojustice education and the revitalization of the cultural commons in San Antonio, Texas and Boone, NC. She also recently served as an International Summer Scholar-in-Residence in Education at Shaanxi Normal University in XiAn, China.

Susan Huddleston Edgerton is Professor of Interdisciplinary Studies at Massachusetts College of Liberal Arts where she teaches curriculum theory, the nature of human nature, introduction to cross-cultural and social justice studies, and environmental justice. She was a professor in the education department from 2004–2015 and director or co-director of the MCLA honors program from Fall 2009 through Spring 2018. Edgerton previously taught at University of Illinois-Chicago and Western Michigan University. She has also been a visiting professor at the University of British Columbia and at York University. She is author of *Translating the Curriculum: Multiculturalism into Cultural Studies* (1996) which received an American Educational Studies Association "Critics Choice" award for 1997, co-editor with Gunilla Holm, Toby Daspit, and Paul Farber of *Imagining the Academy: Higher Education and Popular Culture* (2005), and author of numerous book chapters and journal articles. She served as Secretary to Division B Curriculum for the American Educational Research Association from 1999–2001 and was a co-organizer and first program chair for the Curriculum and Pedagogy Conference. Current interests include environmental sustainability and justice, education, and reclaiming the commons.

Jeff Edmundson was a high school social studies teacher for 24 years. During that time, he also worked in teacher education while completing his dissertation with Chet Bowers at Portland State University. He moved to the University of Oregon in 2008 where he finished his teaching career as Director of Master's Programs and helped create an entirely new teacher education program. The focus of his work has been ecojustice, both in developing the theory by extending it with the pedagogy of responsibility as well as working out the practical application of ecojustice within teacher education. He is co-author with Rebecca Martusewicz and John Lupinacci of the book *Ecojustice Education: Toward Diverse, Democratic and Sustainable Communities, Third Edition*.

David J. Flinders, Professor Emeritus at Indiana University, Bloomington, received his PhD from Stanford University in 1987. He is the co-editor of *The Curriculum Studies Reader, 5th Edition* (Routledge), and has co-authored two books with Chet Bowers, one book with Nel Noddings, and one with P. Bruce Uhrmacher and Christy Moroye. Flinders has also served in a number of elected positions for the American Educational Research Association, including as Vice President (Division B: Curriculum Studies). He recently served as the Curriculum Subject Editor for the forthcoming *Routledge Encyclopedia of Education*. Flinders' professional interests focus on curriculum theory, the cultural ecology of schooling, and qualitative research methods.

Per Ingvar Haukeland is professor of Outdoor Life (friluftsliv) studies with a focus on ecophilosophy and pedagogy at the University of South-Eastern Norway. He holds a PhD from UC Berkeley in Social and Cultural Studies in education. He has collaborated with the Norwegian philosopher, Arne Naess, for 25 years, including on two books: Life's Philosophy (2003) and Deep Joy: Into the depths of deep ecology (2008; in Norwegian). Haukeland was a master's student of Chet Bowers at University of Oregon from 1989–1991. His current research interests are in phenomenology of intra-play, storied wanderings, wayfaring, action-research for sustainability and becoming animal.

Kathleen Kesson is Professor Emerita of Teaching, Learning and Leadership at LIU-Brooklyn. She has written extensively in books and academic journals about democracy and education, teacher development, spirituality and the curriculum, unschooling, environmental education, arts in education, and educational futures. As an environmental activist turned education professor in the 1980s, she credits Chet Bowers with encouraging her to stay true to her experience and sensibilities. Bowers' scholarship continues to inspire her work with Vermont educators and activists dedicated to the transformation of schools and communities to be more sustainable, equitable, just, and joyful spaces.

Andrejs Kulnieks is an Assistant Professor with the University of Saskatchewan in the Department of Curriculum Studies teaching in the area of Secondary School Language Arts, Literacies, and Drama education. He has worked with Undergraduate and Graduate Programs in Education at York University, Nipissing University, and Brock University. His research interests include curriculum theory, language arts, (eco)literacies, arts-informed research, poetic inquiry, Indigenous environmental studies, eco-justice, and environmental education. Kulnieks works from an ecojustice framework to investigate how a deep analysis of language can foster the development of deep relationships with place. In his research about eco-literacy, he explores how oral and literary tradition can enhance and inspire writing practices.

John Lupinacci is an Associate Professor of Cultural Studies and Social Thought in Education at Washington State University. He teaches pre-service teachers and graduate students in the Cultural Studies and Social Thought in Education (CSSTE) program using an approach that advocates for the development of scholar-activist educators. His work as a high school math and science teacher, an outdoor environmental educator, and a community activist all contribute to examining the relationships between schools and the reproduction of the cultural roots of social suffering and environmental degradation. He has co-edited four books, co-authored one book, co-edited four special issues for journals, and is author and co-author of over 40 chapters and articles. He has presented over 100 papers at national and international conferences, including keynote speaking, invited lectures, and workshops. Lupinacci is the past chair of the AERA Environmental Education SIG, and on the editorial board for several leading journals on critical education related to ecocritical pedagogies.

Rebecca A. Martusewicz is Professor Emerita in the Department of Teacher Education at Eastern Michigan University, and Docent Professor at Tampere University, in Tampere, Finland. For 31 years she taught courses within the Social Foundations of Education, focusing specifically on Eco-Justice Education, Globalization, Eco-Feminism, and Advanced Qualitative Methods. She is author of *A Pedagogy of Responsibility: Wendell Berry for EcoJustice Education* (Routledge, 2018), co-author (with Jeff Edmundson and Johnny Lupinacci) of *EcoJustice Education: Toward Diverse, Democratic and Sustainable Communities, Third Edition* (Routledge 2020), and many journal articles and book chapters focused on Eco-Justice Education. She is also author of *Seeking Passage: Post-Structuralism, Pedagogy, Ethics* (Teachers College Press 2001). In the fall of 2015, she was awarded a Fulbright Fellowship to teach and do research around themes of social, political and ecological crises and our responsibilities as educators to address these intersecting problems. A long time collaboration with Chet Bowers early in the development of her work remains foundational.

Pat O'Riley has taught at universities in Canada, New Zealand and the USA, in remote northern communities, First Nations reserves and the High Amazon of Peru. Her teaching is shaped by ecological, Indigenous, poststructural and posthumantist theory and practices. Her research focusses on the intersections of social justice, ecology, technologies and global capitalism. For two decades O'Riely has been conducting research with Indigenous communities in BC in the regeneration of their traditional ecological knowledges and practices. Concerned with global warming and global social inequities, O'Riely's research has expanded to include research with Kichwa-Lamista communities in the High Amazon of Peru to support them in their efforts toward community and ecologial sustainability and to examine the significance of their traditional ecological knowing as counter-narratives to the predominant "progress" narratives offered in mainstream education.

Jennifer S. Thom is Associate Professor in Curriculum Studies and Mathematics Education at the University of Victoria, Canada. An experienced public-school teacher and active community member, she seeks to enable culturally and environmentally generative ways in which to live and learn. Her current research and teaching focuses on explicating (em)bodied and ecologically responsive praxes that re-cognize the live(d) curriculum and STEM with/as place. She is grateful to Chet Bowers and his theory which continue to deeply inform her work, shaping her perceptions, interrupting her thinking, and reminding her of the importance to always be (re)examining taken for granted cultural assumptions.

Kelly Young is a Professor at Trent University's School of Education and Professional Learning where she teaches English curriculum methods and foundational courses. Her research interests include language and literacy, curriculum theorizing, leadership in eco-justice environmental education, and arts-informed research.

Foreword

by William F. Pinar, Series Editor

"How can we account for intellectual breakthrough?" That is the question Charles David Axelrod (1979, 67) asks in his study of Freud, Simmel, and Buber, a question that occurs to me as I reflect on the intellectual breakthrough of C. A. Bowers. Bowers was among the first and most persistent of those demanding that institutionalized education address the intensifying environmental crisis and our complicity in it. Schooling was the culprit in his still stunning comprehensive critique; in Bowers' scholarship schooling becomes almost a synecdoche for how we human beings think and act, explaining what in us human beings caused the calamity occurring right before our eyes. That collage of concerns—ecological, educational, cultural, technological—constitutes the courageous contribution of C. A. Bowers, an intellectual breakthrough at once enabled and resisted by colleagues and students, an accomplishment I attribute to Bowers' individual courage, commitment, and character.

Those qualities could make Bowers contentious; after all, he did not hesitate to criticize those complicit in the ongoing crisis. Constructivism, critical pedagogy, and technological rationality were among the modes and topics of thought and action Bowers accused of complicity in our ecological and civilizational crisis. He outed their arrogant anthropocentricity, condemned their self-righteous insularity, their unself-conscious collaboration in the decimation of life on earth. On one occasion I joined him (Bowers & Pinar, 1992). Searing such sacred cows as constructivism, critical pedagogy, and technologization cost Bowers community—hardly completely, as this collection testifies – but Bowers seemed unconcerned. He knew that:

> Intellectual institutions (like the university) in our society that claim to understand and speak for Western intellectual tradition have often lost the very spirit of that tradition because of their self-limiting orientations. Their structuring of thought has made it inconsistent with thinking itself. Members of these institutions, who have submitted to institutional rule, have become merely its instruments. But more, they have violated the tradition from which they

pretend to speak. The monumental moments of that tradition were certainly not moments of mimesis. They were ones of reflection, critique, and discourse.

(Axelrod, 1979, p. 68)

Axelrod is thinking of Freud, Simmel, and Buber as exemplary instances of this now-defiled Western tradition, individuals whose thought did not succumb to institutionalization, who enacted in their work the ideals of that tradition. Like those three figures, I suggest that C. A. Bowers also "represent[s] the heart of that tradition" (1979, p. 68), an intellectual *engagé*, committed to contest what caused Western cultural ideals to become so defiled and, in so doing, breakthrough those conceptual obstacles blocking their realization.

Of those three "monumental moments" of the Western tradition Axelrod names—"reflection, critique, and discourse"—in Bowers' breakthrough critique is key. It is the central element binding the three together. Bowers (2016, p. 34) criticized colleagues who, he judged, were submerged in simplistic confidence that progress is inevitable: "This hubris is especially prominent in the thinking of professors of education who view their missionary role as ensuring that students march to the current drumbeat of progress, which now requires or reliance upon computer-mediated learning." Bowers' criticism focused Wapner's worry that: "Extreme confidence in human ingenuity and technological prowess, and the faith that humanity is the be all and end all of life on earth, suggests that, if we want, we can bioengineer new species and someday even bring back extinct ones" (2010, p. 154). The hubris that concerned both men is evident in current efforts at "solar climate intervention or solar geoengineering" (Flavelle, 2020, October 29, B3). The crises capitalism creates are those its Frankenstein—technology—can presumably solve.

As Bowers knew, such "extreme confidence"—hubris is his more pointed and precise designation – derives from as it propels those technologies in which we are now so entirely embedded. Bowers was clear concerning his own embeddedness: "Like the technology of print, I am dependent upon using the Internet even as I am aware of its limitations" (2016, p. xv). He suspected that "both print and data represent only a surface account of the emergent, relational, and interactive nature of embodied human experience" (2016, p. 23). While appreciative of Bowers' concerns—his focus on data(ism) was prescient, as recent scholarship confirms (see, for example, Couldry & Mejias, 2019; Koopman, 2019; Williamson, 2017)—I am compelled to point out that being stuck on the surface depends on what is printed, as print can also be a portal to what is underneath the social surface, as the fiction of Virginia Woolf demonstrates. Bowers himself relied on print, and his print provided no "superficial account." Print permeated with presence (Gumbrecht,

2004)—often associated with orality, in Bowers' case with critique—is what enables us to slip below the surface of the social.

Like Indigenous Peoples and non-Indigenous others like the Canadian political economist and communications theorist Harold Innis - Bowers (2016, p. 39) knew that: "Oral cultures also have a more complex understanding of the importance of the non-monetized cultural commons". With the present in shambles—"public space now threatened with extinction by images and simulacra of reality" (Jay, 1993, p. 74)—and the future foreclosed, it is to the past we turn to reactivate our resolve, from where we can critique, engage in reflection and devise discourse, those the three elements of intellectual breakthrough Axelrod identified.

The "questions at the heart of the concept of 'breakthrough' [are] questions about thinking, individuality, and community," Axelrod (1979, p. 2) concludes. As this collection testifies, Bowers contested and created community as he provoked reflection, animated scholarly discourse, all through his individuated thinking. What strikes me about Bowers, then, is less his rich relation to his community and more his single-minded, indeed individual, devotion to his calling: teaching. In his chapter, David Flinders reminds us that Bowers' teaching demonstrated a strong sense of responsibility to generations not-yet-born by addressing those of us alive today, showing students (for example) how the metaphorical character of language can reproduce discredited cultural assumptions from the past, and with catastrophic consequences.

Critique requires reflection: Bowers urged his fellow educators to encourage a heightened consciousness of everyone's situatedness: culturally, politically, biospherically. Today connectivity implies neither social solidarity nor self-development but an Internet connection. Bowers knew that technologization removes us farther from our organicism, from those fellow forms of life sharing the planet we all inhabit. "Marginalized by the digital revolution," Bowers (2016, p. xiv) appreciated, "are the forms of face-to-face, tacit, experience-based, and intergenerational knowledge rooted in local cultures, context, and ethnic practices." Marginalized may no longer be strong enough a term, as digitality destroys culture, updated forms of cultural genocide that Indigenous peoples have suffered for centuries (Dickason & Newbigging, 2010, p. 334; Conn, 2004, p. 30; Hoxie, 2001, p. 128). Replacing embodied place-based oral cultures is cyberculture, a pseudo-culture called the Cloud, e.g., no place in particular.

Technology has long been the accomplice of capitalism, creating pseudo-cultures of consumption and commodification, crafting a commons cut off from actual connection to the natural world, cut off from that cultural patterning (quoting Flinders in Chapter 1) "that connect us with one another and with the natural systems upon which we depend." Once connected but now cut off, "there is a problem of whether there will be any place for different cultural traditions of wisdom and moral values as data, the decision-making programs of computer experts, and the Western myth of the autonomous (that is, self-centered) individual become more widespread" (Bowers, 2016, p. 60). While spot-on, here Bowers' insight also sidesteps the fact that he was himself very much the individual(ist), exercising old-fashioned human—not the self-centered narcissistic sort he is decrying above - autonomy to sound the alarm, over and over again, alerting us to just how wrong our relationship to the natural world has gone.

Curriculum could conceivably come to the rescue, if only curriculum were encoded with "ecological intelligence" (Bowers, 2016, p. 82) instead of "innovation" and "entrepreneurialism," if only curriculum were conservative, "aligned with the interdependent and largely non-monetized world of the cultural commons where the analogs are framed by an awareness of the need to conserve species, habitats, and the gift economy of intergenerational knowledge and kills practices across a wide range of human activity"(Bowers, 2016, p. 81). I, too, have reclaimed the concept of "conservative" from those reckless revolutionaries currently claiming that concept: "To be progressive today is to become conservative, committed to the preservation of public education and, through education, the preservation of the planet" (Pinar, 2019, p. 126). The scale of the curricular challenge is as daunting as it is urgent; as Flinders emphasizes: "Taken together, the chapters in this volume represent not a retrospective of Bowers' work, but only a beginning to the vast work in curriculum studies needed to reorient education in ways that align learning with today's environmental challenges" (quoted from this volume).

There is a subjective side to such a reorientation - such a reconceptualization—one that Bowers (2016, p. 74) himself appreciated: "Neither can the digital revolution lead to the inner transformations in consciousness and self-identity that would lead to adopting a life of voluntary simplicity—to cite just one example of a possible personal transformation." There is in this insight an echo of another great curriculum theorist Dwayne Huebner (1999, p. 408) knew that: "Priority must be given to human beings and the natural order. Then we can see more clearly how humankind participates in the continual creation of the world. We can also see how the "creations" of humankind sometimes bring us closer to extinction." Among those "creations" that bring us closer to "extinction" is technology itself. Bowers (2016, p. 74) knew that digital culture does not enable people to experience the deepest levels of meaning and personal commitment previously associated with the wisdom traditions summarized here. But it can lead to the form of consciousness that reflects the adolescent stage of development promoted by corporate capitalism where everything is exciting, continually changing, free of long-term consequences, and seemingly in endless abundance.

In contrast to the illusory "abundance" capitalism creates, there can be a curriculum of actual abundance, as Jardine, Clifford, and Friesan (2006) affirm, curriculum encouraging what Bowers attests, those "deepest levels of meaning and personal commitment."

Bowers knew we won't see such a curriculum anytime soon, not in time to save the species. James B. Macdonald (1995, p. 73) knew too, emphasizing the apparently impossible choice we face: "Short of detechnologizing society, we are faced with the fact that political action that in any way threatens our fundamental technological cultural base is no longer a viable alternative unless we are willing, in the name of ideals, to inflict untold suffering and the threat of extinction on millions of human beings." Bowers seemed sometimes willing to take that risk, mobilized as he was by his knowledge that the human species - and not only our own, as mass extinction is well underway—is already at risk for extinction. That knowledge animated his tireless teaching.

Bowers' intellectual breakthrough was both paradigmatic—contextualizing the present calamity culturally, conceptually, and technologically, showing how purportedly "progressive" movements like constructivism and critical pedagogy were complicitous—and individual, as Bowers himself interwove insights from several bodies of knowledge to make an original, prolonged, still resounding statement, a teaching taken up so stirringly in this volume. This tireless teacher sought no disciples. Like Georg Simmel—whose accomplishment Axelrod studied – Bowers seemed to say: "I know that I shall die without intellectual heirs, and that is as it should be. My legacy will be like cash, distributed to many heirs, each transforming his part into use according to his nature—a use which will no longer reveal its indebtedness to this heritage" (quoted in Axelrod, 1979, p. 48). That humility steadies us still, we who resolve to remember and enact C. A. Bowers' intellectual breakthrough.

References

Axelrod, C. D. (1979). *Studies in intellectual breakthrough.* Amherst: University of Massachusetts Press.

Bowers, C. A. (2016). *Digital detachment. How computer culture undermines democracy.* New York, NY: Routledge.

Pinar, W. F. & Bowers, C. A. (1992). Politics of Curriculum: Origins, Controversies, and Significance of Critical Perspectives. *Review of Research in Education* (163–190). Washington, DC: American Educational Research Association.

Conn, S. (2004). *History's shadow: Native Americans and historical consciousness in the nineteenth century.* Chicago: University of Chicago Press.

Couldry, N., & Mejias, U. A. (2019). *The costs of connection. How data is colonizing human life and appropriating it for capitalism.* Sanford, CA: Stanford University Press.

Dickason, O. P., & Newbigging, W. (2010). *A concise history of Canada's First Nations* (2nd ed.). Don Mills, ON: Oxford University Press.

Flavelle, C. (2020, October 29). Pileup of climate calamities seeds a radical idea. The New York Times, CLXX, No. 58, 861, B3.

Gumbrecht, H. U. (2004). *Production of presence. What meaning cannot convey.* Stanford, CA: Stanford University Press.

Hoxie, F. E. (2001 [1984]). *A final promise: The campaign to assimilate the Indians, 1880-1920.* Lincoln: University of Nebraska Press.

Huebner, D. E. (1999). *The lure of the transcendent.* Mahwah, NJ: Lawrence Erlbaum.

Jardine, D., Clifford, P., & Friesen, S. (Eds.). (2006). *Curriculum in abundance.* Mahwah, NJ: Lawrence Erlbaum.

Jay, M. (1993). *Force fields: Between intellectual history and cultural critique.* New York, NY: Routledge.

Koopman, C. (2019). *How we became our data: A genealogy of the informational person.* Chicago, IL: University of Chicago Press.

Macdonald, B. J. (Ed.). (1995). *Theory as a prayerful act: The collected essays of James B. Macdonald.* New York, NY: Peter Lang.

Pinar, W. F. (2019. *What is curriculum theory?* (3rd ed.) New York: Routledge.

Wapner, P. (2010). *Living through the end of nature: The future of American environmentalism.* Cambridge, MA: The MIT Press.

Williamson, B. (2017). *Big data in education. The digital future of learning, policy and practice.* London: Sage.

1 An Introduction to the Curriculum and Environmental Scholarship of C. A. (Chet) Bowers

David J. Flinders

The touchstone of this volume is the educational and environmental scholarship of C. A. (Chet) Bowers (1935–2017). Bowers was a prolific writer across topics ranging from curriculum theory and school reform to ecological literacy, the cultural roots of modernity, sustainability, and the uses of digital technologies. Across his career, Bowers authored 27 books and close to 200 journal articles and book chapters. His work has been translated into Spanish, Chinese, and Japanese. In addition, Bowers served on international commissions and actively promoted his work through extensive speaking engagements. He spoke at the invitation of 42 universities across the United States, and 41 universities in other parts of the world, including England, Canada, Switzerland, Ireland, China, South America, Australia, South Korea, Taiwan, and Japan. As a consequence, Bower's pioneering contributions to curriculum and ecological studies have inspired other scholars and continue to hold far-reaching significance in these fields. As Pinar, Reynolds, Slattery, and Taubman (1995) put this, "Bowers' scholarship is significant, and careful, respectful attention to it by the curriculum field is long overdue" (p. 271).

The power of Bowers' scholarship is found in its overall coherence and cogency by which he brought together perspectives from a range of disciplines, including philosophy, sociolinguistics, anthropology, sociology, and social criticism. While its expanse makes this work difficult to summarize in any brief format, the broad contours of his work, the ideas that animated Bowers' labor, can be traced back to his graduate studies in the intellectual traditions of Western thought. With the Enlightenment as a salient point of departure, Bowers painstakingly established clear through-lines from writers such as Rene Descartes, Francis Bacon, and Johannes Kepler to the ecological crises we now face in the 21st Century. Drawing from this history, Bowers critiques cultural frameworks and patterns of belief that have propelled the human degradation of the earth's natural systems.

From Enlightenment thought, Bowers identified a network of root metaphors that today persist as cultural anchors. Root metaphors are words or phrases that bring with them patterns of thought established in the past. These metaphors include change as progress, individuals as autonomous, nature as a machine, technology as neutral, and language as an ahistorical tool. Change as progress, for example, assumes the ascent of modernity as promised by the scientific revolution. In most Western cultures, this ascent is represented by mass consumption and today's rapid depletion of natural resources.

Individuals as autonomous is another example of a root metaphor that holds far-reaching implications. Here individuals are represented as the nucleus of reflective thought and independent actions. Bowers rejected this metaphor on the basis that it removes individuals from the powerful but often unrecognized influence of both language and culture. Bower was not alone in his critique. Rather, the autonomous individual has been questioned in ongoing scholarship by feminist writers as an illusion that legitimates patriarchy, by behavioral economists as a poor predictor of human action, and by cognitive scientists as neglecting assumed patterns of thought inherited from the past. Such ongoing work signals the possibility of cultural change.

Environmentalism

Across his career, Bowers thinking became increasingly focused on the cultural dimensions of today's ecological crisis. In the 1970s he drew on the sociology of knowledge, a field that made the concepts of culture and socialization central to understanding the social construction of reality and the influence of prevailing ideas on social institutions. For Bowers, culture represented the templates of shared knowledge, skills, values, understandings of time and space, traditions, and material objects of a group. In developing this cultural perspective, Bowers drew on anthropologists and particularly Gregory Bateson's (1972) book, *Steps to an Ecology of Mind*. While we often think of natural ecologies as comprising land, plants, animals, and insects, Bateson's ecology of mind is populated by relationships, cultural norms, shared ideas, traditions, and language. Such cultural ecologies, again like natural ecologies, are primarily characterized by looped systems of information exchange and transformation. Ecologies are environments rich in information; so rich that it becomes impossible to separate off components or "individuals" from the systems of which they are a part. By way of contrast, traditions in psychotherapy have sought to extend our understandings of the mind inwards by delving into subconscious systems such as the id, ego, and super-ego. Bateson, a semiotic anthropologist, extended the mind outward as interdependent and fully integrated with natural, cultural, and social environments.

Drawing from Bateson's work, Bowers relied on the fundamental principle that all of the components of information exchange form an ecology dependent on one another. To put this another way, none of the components of an ecology (natural or cultural) is capable of exercising unilateral control over any of the ecology's other components. Thus, an ecological framework asks that we learn to understand causality as circular rather than linear. A wall-mounted thermostat in a house, to take a simple and commonplace example, is part of a system that typically includes a furnace (or AC) as well as other components. Yet we describe the function of thermostats in revealing ways. We say that the thermostat controls the furnace, which in turn controls the temperature inside the house. From an ecological framework of complex looped systems, the temperature "controls" the thermostat, which controls the furnace, which "controls" the temperature. I put the words control in quotation marks because no component of an ecology has unilateral control over the system as a whole. That is, the components of an ecology are not so much controlling as they are responsive and interdependent.

Moreover, we can add other systems to the thermostat-furnace-temperature loop. We can extend the circuit be including the self or the power plant that supplies energy. Individual people and power plants are also systems, dependent on other systems. When we include individuals in the system, they too are typically responsive. If the temperature in a room becomes too hot or too cold, the occupant is likely to get up and adjust the thermostat.

Bowers was further influenced by a range of environmental writers. Key figures include deep ecologists such as Wendel Berry and Aldo Leopold as well as ecojustice authors such as Vandana Shiva and Gary Snyder. Such writers have persistently stressed the interdependence of natural and cultural systems. Like Bateson, moreover, ecojustice authors embrace this interdependence not simply as a biological principle, but as an ideological and moral stance. When an individual or a group acts on the basis of self-interest, that which concerns me or my particular group, they enclose themselves as separate and cut off from other systems. They decide to rid their waste in ways that degrade natural systems of which we are a part. Bateson (1972) puts this principle in epistemological terms: "... while I can know nothing about any individual thing by itself, I can know something about the relations between things" (p. 157).

Language and Thought

One of Bowers' central concerns was that ecological thinking is largely put out of focus or undermined by the very language of Western cultures. Bowers (2018) referred to the influence of language as the linguistic roots of the ecological crisis, noting that, "There is, in short, no area of the curriculum that does not rely upon the language systems of the culture, or that of other cultures" (p. 43). In everyday circumstances, we often think of language as a highly flexible tool, a conduit for transmitting ideas among groups and individuals. This conduit view of language is deceptively simplistic because it suggests that our thoughts influence what we say, but not vice versa. Bowers' position holds that words are neither neutral nor passive in their relationship with thought processes. On the contrary, because words have a history, they reproduce distinct cultural patterns of thought from the past. As Bowers (2018) puts it, "… acquiring the language of one's community also involves being dependent upon ways of thinking about issues and problems that were unknown in earlier times" (p. 47). As we speak language, language speaks us (Heidegger, 1927/1962).

In large part, this influence is achieved through the metaphorical nature of language. The word metaphor comes from the Greek, *metaphora*, meaning "to carry over." That is, metaphorical language is a way of understanding something in terms of something else. Analogic metaphors typically include a source domain based on familiar experiences and a target domain representing something that is new or unfamiliar. While such metaphors bring into focus the similarities between their source and target domains, they do so only by putting out of focus the domains' differences. By bringing forward certain characteristics of an experience but not others, metaphors express an inherent point of view. As the poet Robert Frost counseled: "All metaphor breaks down somewhere…. It is touch and go with the metaphor, and until you have lived with it long enough you don't know where it is going. You don't know how much you can get out of it and when it will cease to yield" (Cox & Lathem, 1968, p. 33).

Bowers' interest in metaphorical language is cultural rather than poetic, and cultural metaphors are pervasive throughout our everyday speech and writing. That is, metaphors do not simply adorn language, but are ways of understanding the new in terms of the familiar. In education alone we speak of tracking, kindergarten and high school, covering curriculum content, homework, motivation, educational aims, individual achievement, school boards, accountability, constructivism, block scheduling, achievement gaps, and benchmarks—examples that only begin a long list of common metaphors that we rely upon when speaking and writing about school experiences.

Curricular subject areas are also rife with metaphorical language. Science teachers talk about genetic engineering, DNA codes, and how genes "behave." Again, such metaphors do more than describe; they also reproduce points of view. An American history teacher or text, to take another example, may devote lessons to the topic of "Westward Expansion." The familiarity with the concept of expansion is assumed; balloons expand, waistbands expand, and so on. Nevertheless, Westward Expansion makes sense for describing the experience of certain groups but not others. If you happen to be on the eastern side of this frontier, one's field of experience and action is expanding. Yet those on the western side do not experience expansion; they experience contraction or encroachment. This is one place where, as Frost put it, the metaphor "breaks down."

Metaphors such as Westward Expansion and genetic engineering can be made explicit, as I have done briefly here, but doing so is an exception that proves the rule. In our everyday talk, we take such metaphors for granted, aware of their influence on thought only at an implicit level of understanding. As Martin Heidegger (1927/1962) puts it, "... language already hides in itself a developed way of conceiving" (p. 199). Metaphorical language, in short, constitutes cultural blinders of which we are unaware. This lack of awareness is due in large part to the ubiquitous nature of understanding the new in terms of the familiar. Much like culture at large, language is too much with us, or too much a part of us to be readily recognized.

Nevertheless, to make metaphors explicit, especially the root metaphors that concern Bowers, is fundamental to the teacher's role in primary socialization and in mediating cultural patterns of thought. For Bowers, teaching includes a responsibility to help students better understand the metaphorical nature of language and how it reproduces cultural frameworks from the past. Yet he also wanted teachers to facilitate a heightened awareness of the implied patterns of understanding that connect us with one another and with the natural systems upon which we depend. These "patterns that connect" (Bateson, 1979, p. 8) are represented foremost in what Bowers called the "cultural commons." The cultural commons include all of the shared community knowledge on which that community draws it functions and identity. This knowledge often includes arts and crafts, face-to-face mentoring, ways of growing and preparing foods, how to perform rituals or rear children, forms of recreation, and so on. The cultural commons will be discussed in detail later in this volume. My point here, however, is that throughout his career, Bowers was reform-minded and committed to finding ways forward.

Closely related to this responsibility, Bowers highly valued the knowledge of practitioners and others steeped in the practical affairs of K-12 public schooling. Without question, Bowers was foremost an academic, embracing and embraced by theory. But Bowers consistently reached out to classroom teachers and school administrators. I believe this is one reason that in 1989, Bowers invited me to co-author a book with him on the ecology of the classroom. I had recently joined the University of Oregon's faculty as a classroom-based researcher, and like so many of the graduate students that Bower's mentored, I was spending a lot of my time in public schools. Bowers welcomed and respected this side of my work as complementary to his own. Our collaboration yielded two books, and over the next 28 years, Bowers would serve as an extraordinary mentor and friend.

Overview of What is to Come

The purpose of this volume is to examine and build on Bowers' ecological approaches to curriculum theory, design, and evaluation. Bowers advocated for school curriculum that supported sustainability, ecojustice, and ecological intelligence. In order to address these issues in depth, we have recruited authors from a pool of international scholars whose own work builds upon Bower's ecological perspectives. Many of these authors also worked directly with Bowers as students or colleagues, offering insights into ecological thinking drawn from a range of perspectives. We have organized their chapters into three parts. Each part of the book includes three or four chapters and is introduced through excerpts drawn from Bowers' previously published work. The excerpts for each part serve two functions. First, we wanted to bring Bowers' voice directly into the book, and the excerpts from his previous work allow for his presence in word and spirit. Second, the excerpts serve to orient the chapters in each of the three parts. Early in the process of recruiting contributing authors, we provided each with the excerpts for their sections, and without didactic instruction, we simply asked authors to read, engage with, and where possible, integrate or build on ideas drawn from the excerpts.

Part I of this volume, *Ecological Approaches to Curriculum Discourse*, includes four chapters each examining ecological literacy in international settings. Each chapter reflects Bowers' engagement with key theorists, picks up where their dialogues left off, and pushes further salient ideas to signal new perspectives. Part I begins with an excerpt from Bowers' (1996) article, "The Cultural Dimensions of Ecological Literacy." In this essay, Bowers grapples the central question of why it has been so difficult for modern societies to learn and practice ecological literacy. In response to this question, Bowers argues that the scope of environmental education be expanded from liberal ideologies to include cultural/ bioconservatism. The term cultural/bioconservatism characterizes traditional cultures that have developed more sustainable knowledge and practices than have modern cultures. Bowers was fully aware of the dangers of romanticizing traditional cultures, but he also notes key lessons to be learned from such cultures in order to help modern societies become less individualistic, human-centered, and consumer-oriented. Moreover, cultural/bioconservatism holds far-reaching implications for reforming K-12 schools and higher education in ways that address today's environmental challenges.

The first chapter of Part I is eco-scholar Jennifer Thom's "Co/inspiring Ecological Conversations with Chet A. Bowers (1935–2017) and Ted

T. Aoki (1919–2012)." In this chapter, Thom focuses on imagined but potentially constructive exchanges between Bowers and Aoki. These deliberations are set within the context of print, data, and technology to reveal both similarities and differences in how Bowers and Aoki define ecological literacy and the primary role of language in cultural reproduction. In particular, Thom traces Bowers' shift from cultural to ecological literacy back to the early 1970s. Bowers was not optimistic in the possibility of Western cultures adopting ecologically literate practices. Thom, nevertheless, draws on Aoki's key concepts of tensionality, bridge which is not a bridge, and the space in-between to complement Bowers' insights. Generating discursive connections between ecological perspectives and modernity, Thom opens a new theoretical space in which to reconceptualize emergent global challenges.

The second chapter of Part I, "Reconceptualizing the Experiencing Subject for the Anthropocene," is by author and poet Kathleen Kesson. This author recounts being introduced to Bowers' work as a graduate student and progressive activist. Like other contributing authors, this introduction led Kesson to question her underlying assumptions as a learner and teacher. In particular, Bowers' 1987 book, *Elements of a Post-Liberal Theory of Education*, prompted Kesson to revisit those assumptions that underpin the tenets of progressive thinker John Dewey. Beyond critique, however, Kesson finds in Bowers' work alternatives to modern Western narratives. These alternatives include the traditional knowledge and skills of the cultural commons. Using relational ontologies and epistemologies as a conceptual backdrop, Kesson then provides a cogent analysis of Dewey's conception of experience as a central notion to progressive thought. This author recognizes important elements in Dewey's work that allow for its rehabilitation in education for sustainability. She thanks Dewey for his insistence on the organic connections between learning and active experience, but Kesson also pushes notions of experience beyond Dewey's reflective views of rationality to include indigenous knowledge, intuition, esthetics, and spirituality. Early in her chapter, Kesson suggests that future curriculum historians are likely to see Bowers "… as the most misunderstood, misrepresented, provocative, persistent, and prescient educational thinker of the current era" (this volume, p. 45). Overall, Kesson's chapter goes a long way in explaining why this may likely be the case.

The third chapter of Part I is by Jeff Edmundson, a scholar whose own work combines ecojustice pedagogy with insights drawn from critical theory. His chapter, "Beyond the Binary of Bowers," reviews past and ongoing tensions between Freiran critical theorists and Bowers' cultural dimensions of sustainability. He laments that both Bowers and the Freirans often talked passed one another, which Edmundson illustrates with the exchange between Peter McLaren and Donna Houston

(McLaren & Houston, 2004) and Bowers (2005). Edmundson admits that critical pedagogy is often "trapped" by its unquestioned embrace of "freedom" and its poor fit with non-industrial cultures. Eco-pedagogy, for its part, has failed to use the genuine insights of critical theory into the unjust reproduction of racism and classism. Edmundson's efforts to correct this lack of dialogue are represented in a third position partially inclusive of both views. Edmundson and others (Edmundson & Martusewicz, 2013; Martusewicz, Edmundson, & Lupinacci, 2015) call this approach a pedagogy of responsibility. A pedagogy of responsibility accepts the need for freedom from oppressive forces, but such freedom is restrained by our obligations to non-human communities and to our own cultural sustainability.

The final chapter of Part I is by eco-philosopher Per Ingvar Haukeland. Haukeland, like Thom, compares Bowers' work with the work of another scholar, this time that of Norwegian philosopher and founder of the deep ecology movement Arne Naess. As Haukeland notes, both Bowers and Naess, who exchanged views in a 1993 publication (Bowers, 1993b), were devoted to finding solutions to the ongoing ecological degradation of natural systems. Haukeland reviews their exchange by examining how each scholar uses key terms, including eco-culturalism, self-realization, and cultural sustainability. In order to ground these concepts in educational practice, Haukeland further situates Bowers and Naess' ideas in relation to *friluftsliv*, a broad term associated with outdoor or "open air" education in Norway. Haukeland demonstrates that while *friluftsliv* is believed to contribute to sustainability, it is also part of our modern life and thus may unintentionally convey similar assumptions as those that have given rise to our industrial and information-based societies. Haukeland's aim is to rethink these assumptions based on cultural and deep ecology principles as they apply to both rural and urban settings.

Part II of this volume includes three chapters that elaborate on the cultural commons as it relates to curriculum studies at large. It begins with an excerpt from Bowers' 2016 article, "Reforming Higher Education in an Era of Ecological Crisis and Growing Digital Insecurity." In this article, Bowers argues that while industrial and post-industrial societies have led to consumer-dependent lifestyles and the disruption of natural systems, we are not without alternatives. These alternatives, collectively known as the cultural commons, include intergenerational knowledge and skills that are passed on through face-to-face interactions, language, forms of education and mentoring, crafts, customs associated with the cultivation and preparation of food, as well as a wide range of social rituals. Bowers suggests the significance of the cultural commons when he writes, "Awareness of climate change without any sense of the existing cultural alternatives leaves us in the same confused state of powerlessness as our leading politicians who are in denial about the relationship between consumer-dependent industrial culture and environmental degradation" (p. 115). How we come to acquire this awareness of alternatives and how these alternatives are under continual threat of enclosure are examined in Part II.

This first chapter in Part II, titled "On Traditions and the Commons: A Material Feminist Analysis" is written by feminist and curriculum scholar, Audrey M. Dentith. Dentith's chapter examines Edward Shuls' concept of tradition, one of the central scholarly inspirations for Bowers' concept of the cultural commons. She simultaneously introduces readers to the work of feminist materialist scholars such as Grosz, Barad, Tuana, and Alaimo and their theories of materiality, ethics, and politics. Building on a feminist materialist frame, Dentith recounts Bowers' call to revitalize the cultural commons, and Grosz' theories of materiality, ethics, and politics. Building on feminist materialism, Dentith develops two key premises that extend materiality to explain human and non-human interdependences. The first premise acknowledges the agency and life force in all physical and other-than-human entities. The second premise is that the physical, social, and cultural worlds are so closely intertwined that they must be considered together. She maintains through her work that there is a blurring of the boundaries between the social/cultural and the physical worlds, a premise that is central to the work of Bowers as well as material feminists. Dentith's approach considers the interconnectivity of the scholarship on feminist materiality, tradition, and the cultural commons in support of new approaches to the curriculum in a time of environmental, political, and social crisis.

The second chapter in Part II, "The Curriculum of the Commons: Chet Bowers in Detroit," is by Rebecca Martusewicz. Martusewicz examines the intersections between social and ecojustice within the context of Detroit's revitalization over the past serval decades. She recounts how shortly after meeting Bowers in 2000, she and Jeff Edmonson organized a series of retreats for educators, community activists, and graduate students. After two successful retreats, the third was held in Detroit in 2004. For this gathering, local activists were invited to describe their revitalization work around community arts, urban agriculture, food banks, youth camps, and mentoring programs, many of which the retreat participants visited. Martusewicz describes "the fundamental lesson" of these programs and of their own work as coming to realize the critical importance of patterns that connect people with each other and with the environments of which they are a part. Citing Winkler-Prins and DeSouza (2005), Martusewicz describes the outcomes of this lesson as "an economy of affection," which adheres to the premise that the pandemic offers us important lessons for the importance of living lives in harmony with all living and non-living entities. She ends with the premise that the pandemic and its residual effects offer a unique opportunity to reimagine curriculum and instruction in the interest of preserving and revitalizing our environmental and cultural commons.

The third chapter in Part II, "Lessons from a Pandemic: Can We Reclaim Our Cultural and Environmental Commons?" is by Susan Huddleston Edgerton. Edgerton situates ecological thought, and particularly Bowers' work, within the context of the 2020 COVID-19 pandemic. She argues that the pandemic has sharply foregrounded the same cultural dimensions of today's ecological challenges that concerned Bowers and continue to worry many curriculum scholars. Writing in the midst of the pandemic, Edgerton contends that social, political, and cultural responses have been complex and at times contradictory. On the one hand, many have responded with denial and mistrust (of experts in particular). The pandemic has also laid bare deep inequalities in its disproportionate impact on the poor and other marginalized groups. On the other hand, sheltering in place has slowed down the routines of daily life and interrupted consumerist behaviors, thus providing opportunities for reflections on the values of a simpler life. Responding to the pandemic, many people volunteered to deliver groceries, donations to foodbanks soared, while others sowed masks and rekindled past relationships. As Edgerton recounts, large numbers of people have faced tremendous hardships due to the pandemic: Loss of loved ones, joblessness, huger, and anxiety about the future. Without discounting such struggles, Edgerton adheres to the premise that the pandemic offers us important lessons for living our lives in harmony with all living and non-living entities. She ends with the premise that the pandemic and its residual effects offer a unique opportunity to reimagine curriculum and instruction in the interest of preserving and revitalizing our environmental and cultural commons.

The final section of the book, Part III, picks up on themes of ecojustice pedagogy and extends them from a variety of perspectives. Part III begins with an excerpt from a 2002 article, "Toward and Eco-Justice Pedagogy," in which Bowers argues that the emerging root metaphors of the environmental sciences are shifting from those that are mechanistic and human-centered to metaphors of relationality and interdependence. The later serve as a conceptual basis for ecojustice education. In particular, Bowers contends that ecojustice education includes three primary foci. The first is on environmental racism and class discrimination. This focus includes knowledge of how today's environmental policies have had a disproportionate impact on poor and otherwise marginalized groups. The second focus is on strengthening non-commodified aspects of local communities. This focus stands in sharp contrast with, in Bowers' (2002) words, "The relentless drive to commoditize more aspect of daily life, and thus create new markets and thus new forms of dependencies ..." (p. 24). The third focus of eco-pedagogy is on our responsibilities to future generations. This responsibility is represented in curricula that aim explicitly at democratizing technology, science, and sustainable practices.

The first chapter in Part III, "Developing Ecological Literacy as a Habit of Mind in Teacher Education through Ecojustice Progressive Curricula," is by Kelly Young. In this chapter, Young expands on the ways in which she brings forth Chet Bowers' ecojustice foundational framework into pre-service teacher education by outlining curricula that promotes Bowers' advocacy for educational reform in terms of reducing "the impact of the industrial/consumer dependent culture on everyday life." Bowers' understanding of the role of language processes and culture in our complex relationships with the natural world help to lay the foundation of ongoing curricular infusion of ecojustice principles into pedagogical practices envisioned for the advancement of ecological frameworks in the field of curriculum studies.

The second chapter of Part III, "Coyote and Raven Encounter Chet Bowers in Conceptual Time-Space: Ecojustice Pedagogies of the Land," is contributed by Peter Cole and Pat O'Riley. As they write, "... we join this conversation on the work of Chet Bowers in the shape of an oral (speaking on the page) narrative that journeys across and between the Global South and Global North." This narrative includes coyote, raven, the ghost of Chet Bowers, and Shakespeare as they pursue a conversation around ecojustice education that goes beyond progressive ideologies to include non-human and more-than-human intelligences. Time is represented as a nuclear doomsday clock; space is represented in languages, the virtual, pedagogies of the land, and the commons. Their narrative interweaves not only Bowers' ideas with indigenous knowledge, but it also suggests Bowers' hopes and frustrations.

The third chapter of Part III, "Developing Relationships with Unfamiliar Places through Poetic Inquiry; A Curriculum of Travel and Song Writing," is by Andrejs Kulnieks. Kulnieks celebrates the commons through poetry, song, travel, and movement. His chapter examines how learners can develop a deeper understanding of intact-ecosystems through engaging in environmental learning over a course of time. In particular, Kulnieks uses his own poems in recounting a journey north to the University of Saskatchewan, where Chet Bowers held his first academic appointment. By juxtaposing such inquiry with understandings of place, Kulnieks demonstrates the importance of local knowledge of, for example, food preparation as a way of strengthening the commons and reducing our environmental impact for the sake of future generations.

The final chapter in Part III is by Johnny Lupinacci. His chapter, "Ecocritical Pedagogies: (Re)Imagining Education for Diversity, Democracy and Sustainability as Eco-Justice Curriculum," addresses the contributions of C. A. Bowers and his conceptualization and introduction of "eco-justice" to widely spread networks of higher education, community activism, and global politics. While contextualizing Bowers' work widely, this chapter focuses on his contribution to eco-critical pedagogies and teacher education—specifically, the importance of curriculum studies in fostering teacher education focused on learning to recognize and rethink often un-identified cultural assumptions embedded in Western industrial culture (Bowers, 1997). The chapter highlights Bowers' work and shares how connected work in curriculum studies supports eco-critical scholars in teacher education.

Taken together, the chapters in this volume represent not a retrospective of Bowers' work, but only a beginning to the vast work in curriculum studies needed to reorient education in ways that align learning with today's environmental challenges. Because these challenges are historically and culturally grounded, they include a wide range of educational opportunities at all levels. These opportunities transverse what Elliot W. Eisner (1992) called the ecology of schooling, an ecology that includes what we teach, why we teach it, how we organize schools, our methods of teaching, and how we evaluate such work on an ongoing basis. As stewards of this cultural ecology, educators hold an inescapable responsibility to the future; the work that follows is a beginning.

References

Bateson, G. (1972). *Steps to an ecology of mind.* New York: Ballantine.

Bateson, G. (1979). *Mind and nature: A necessary unity.* New York: Dutton.

Bowers, C. A. (1987). *Elements of a post-liberal theory of education.* New York: Teachers College Press.

Bowers, C. A. (1993a). *Educating for an ecologically sustainable culture: Rethinking moral education, creativity, intelligence, and other modern orthodoxies.* Albany, NY: State University of New York Press.

Bowers, C. A. (1993b). Some questions about the theoretical foundations of W. Fox's transpersonal ecology and Arne Naess' Ecosophy T. *Trumpeter, 10*(3), 2–16.

Bowers, C. A. (1996). The cultural dimensions of ecological literacy. *Journal of Environmental Education, 27*(2), 5–10.

Bowers, C. A. (1997). *The culture of denial: Why the environmental movement need a strategy for reforming universities and public schools.* Albany, NY: State University of New York Press.

Bowers, C. A. (2002). Toward an eco-justice pedagogy. *Environmental Education Research, 8*(1), 21–34.

Bowers, C. A. (2005). How Peter McLaren and Donna Houston, and other "green" Marxists contribute to the globalization of the West's industrial culture. *Educational Studies, 37*(2), 185–195.

Bowers, C. A. (2016). *Reforming higher education in an era of ecological crisis and growing digital insecurity.* Process Century Press, Anoka: Minnesota.

Bowers, C. A. (2018). *Ideological, cultural, and linguistic roots of educational reforms to address the ecological crisis: The selected works of C. A. (Chet) Bowers.* New York: Routledge.

Cox, M., & Lathem, E. C. (1968). *Selected prose of Robert Frost.* New York: Collier.

Edmundson, J., & Martusewicz, R. (2013). Putting our lives in order: Wendell Barry, ecojustice and a pedagogy of responsibility. In A. Kulnieks, K. Young, & D. Longboat (Eds.), *Contemporary studies in environmental indigenous pedagogies: A curricula of stories and place* (pp. 171–184). Rotterdam, Netherlands: Sense Publishers.

Eisner, E. W. (1992). Educational reform and the ecology of schooling. *Teachers College Record, 93*(4), 610–627.

Heidegger, M. (1927/1962). *Being and time.* New York: Harper & Row.

McLaren, P., & Houston, D. (2004). Revolutionary ecologies: Critical pedagogy and ecosocialism. *Educational Studies, 36*(*1*), 27–44.

Martusewicz, R., Edmundson, J., & Lupinacci, J. (2015). *Ecojustice education: Toward diverse, democratic and sustainable communities.* New York: Routledge.

Pinar, W., Reynolds, W., Slattery, P., & Taubman, P. (1995). *Understanding curriculum: An introduction to the study of historical and contemporary curriculum discourses.* New York: Peter Lang.

Winkler-Prins, A. M. G. A., & DeSouza, P. (2005). Surviving the city: Home gardens and the economy of affection in Brazilian Amazon. *Journal of Latin American Geography, 4*(1), 107–126.

Part I

Ecological Approaches to Curriculum Discourse

David Orr's observation that "[a]ll education is environmental education" (1992, p. 90) is a good place to begin considering the cultural dimensions of ecological literacy. However, understanding the full implications of his insight requires that two possible sources of misunderstanding be addressed. The first is that Orr's statement should not be interpreted to mean that all environmental education contributes to more sustainable cultural/natural systems relationships. Environmental education in certain areas of the curriculum, and in certain sectors of society, may communicate the message that natural systems are incidental to the drama of human life or that they are to be viewed as "natural resources" to be exploited as quickly as consumer markets can be created for them.

That is, environmental education often takes a destructive form. The second source of misunderstanding results from viewing education only as occurring in public schools or universities. To understand the complexity of the educational processes that contribute to various and often contradictory forms of environmental education, it is necessary to recognize that education, in the broadest sense, is synonymous with culture. The cultural patterns that give predictability to everyday life are sustained and renewed through communication. The multiple language systems that reproduce these cultural patterns as individuals interact with each other and the natural environment are part of the ongoing process of education. Culture, in effect, represents earlier ways of understanding that are encoded: (a) In material objects such as the design of cars, buildings, and computers, (b) at the level of taken for granted patterns of interaction and thought, and (c) even in the intentional and reflective interpretations of everyday life. As the cultural form of intelligence is communicated through these language systems, new members learn the old patterns—often at a taken for granted level of understanding.

Thus, I can restate Orr's insight in the following way: "All forms of communication essential to sustaining cultural patterns are part of the process of environmental education." Ecological literacy, which too often is associated only with activities that occur in schools, can now be broadened to include an awareness of how the assumptions, values, technologies, and categories of thinking of a culture influence how humans relate to the environment.

The complexity of the challenge of fostering a form of environmental education that contributes to cultural/ecosystem sustainability be seen when we consider how the deep cultural assumptions that underlie the consumer-technological mainstream culture are reinforced by the liberal ideologies that both guide and legitimate the direction educational reform has taken over the last 100 years—and which continue to insure that the more destructive forms of environmental education will be perpetuated by the educational reform efforts we are now witnessing. These deep cultural assumptions include the following: (a) That change and experimentation with the foundations of culture are inherently progressive in nature, (b) that an anthropocentric interpretation of the Earth's ecosystems represents the highest expression of enlightened thinking, (c) that individual autonomy in the areas of thought and values represents the fullest realization of human potential, (d) that science and technology are the twin engines of human progress, and (e) that the Western form of modernization represents the most advanced stage of human development and should be promoted throughout the world. Aside from recent conservative" critiques, which really represent an attempt to conserve educational traditions that were the basis of the Industrial Revolution, educational reform in public schools and universities continues to be driven by liberal ideologies that reinforce these ecologically unsustainable assumptions.

The reason that the various traditions of liberalism have not led educators to consider the cultural roots of the ecological crisis becomes clearer when we sort out what is distinctive to each of the traditions of liberalism, as well as what is shared in common. The shared assumptions often are not associated with the three distinctive educational expressions of liberalism (identified here as technocratic, romantic, and emancipatory liberalism) because their distinctive emphasis has become the basis of how we identify them. In effect, that is distinctive to each position is what has been foregrounded in their public and professional discourse.

Technocratic liberals, for example, view scientifically-based, procedural problem-solving forms of knowledge as leading to engineered environments that will shape behavior in predictable ways. Their concerns with engineering educational environments range from behavioral management in the classroom to creating a more functional fit between the educational process and the needs of the ever changing workplace. With the development of computer-based technologies, technocratic liberals are now adopting a key assumption of romantic liberals—that individuals construct their own ideas on the basis of data—which computers make available on a massive scale. The romantic liberals in the early 1900s, and later in the 1960s and early 1970s, argued that students create their own ideas and values from direct experience. Ironically, the romantic liberals have recently adopted the technocratic liberals' way of reducing experience, including all of its relational and contextual elements, to data. The emancipatory liberals, who have a long history of urging educational reforms that lead to a more egalitarian and democratic society, continue to emphasize the importance of fostering critical inquiry in the classroom. In spite of these differences between the different forms of liberalism, the irony remains that educational reform continues to be based on the same cultural assumptions that gave legitimacy to the Industrial Revolution and, now, the emerging Information Age. It should also be emphasized that the form of environmental education consistent with these cultural assumptions has little resemblance to the form of environmental education that occurs in cultures that have evolved along more ecologically centered pathways.

The recognition that many traditional cultures developed highly complex patterns for insuring that the wisdom accumulated and refined over generations of dialogue between humans and their environment was passed on to future generations leads us back to the question: "Why has ecological literacy become so difficult for modern cultures to understand and carry out successfully?" Part of the answer, I would suggest, can be found in how the early Western mythopoetic narratives represented humans as separate from nature—as being in control of their own destiny regardless of how their actions degraded the environment. Another part of the answer can be found in the modern practice of nearly every aspect of the human/natural world—thus extending the market system, with its emphasis on economic values and technological innovations. A third reason can be found in the increasing dominance of science as an explanatory framework. Although science has added to our understanding of natural phenomena and to our technological prowess, it represents a discourse that acknowledges that moral values are beyond its legitimate domain of inquiry. At the same time, science undermines the authority of the metanarratives that are the foundation of our moral codes. A fourth reason that ecologically sustainable environmental education has eluded modern cultures can be traced back to the dominant ideologies that play the same role in modern cultures that mythopoetic narratives play in traditional cultures. Educators are not in a position that allows them to alter the influence of the Judeo-Christian creation stories, nor do they possess the ability to alter in any significant way the continuing commodification of life forms (biotechnology being the latest frontier). The secularizing influence of science is also beyond their domain of influence. But they can challenge the ideological/epistemological orientations that make transgenerational communication so problematic in modern cultures.

Although I am aware of the danger of romanticizing traditional cultures, and of equally significant problems associated with borrowing ecologically sustainable patterns from other cultures, I think we can learn from cultural groups who have evolved highly complex symbolic worlds that are relatively sustainable. One of the most important lessons educators can learn from these cultures is that the successful ones all adhere to belief/moral/technological systems that can best be described as having a cultural/bioconservative ideological orientation—to use a modern term.

The central point about cultural/bioconservatism is that it conserves and renews cultural patterns that minimize the adverse impact of humans on natural systems. By recognizing that cultures must be judged in terms of their impact on the diversity and viability of ecosystems, a different form of accountability is being introduced. Thus, cultural/bioconservatism is not a form of conservatism that maintains the privileges of certain groups or conserves traditions out of nostalgia for the past. Rather, it represents an expansion on Aldo Leopold's land ethic: "A thing is right when it tends to preserve the integrity, stability, and beauty of the biotic community. It is wrong when it tends otherwise" (1966, p. 262). By framing his land ethic in the more comprehensive language of modern ideology—that is, its moral framework, the forms of knowledge that are privileged, and the kinds of relationships that are the basis of daily life—we avoid a problem that Leopold did not recognize. Leopold framed the land ethic as a guide for rational individuals, but the recognition that rational individuals often cannot transcend self-interest, or understand the common good in the same way, leads to a cultural rather than an individually-centered approach. The shift to a cultural perspective also brings us back to a point made earlier: namely, that all the language systems of a culture frame how relationships are to be understood and negotiated. These relationships, whether they occur within the family, between strangers, as part of business activity, or between humans and the natural environment, always involve the moral question: How should think and act in this relationship? Given that these languages are cultural, that is, encode earlier shared understanding and patterns, they reproduce the moral templates that are to be used in different relationships.

How would an educational approach to ecological literacy that is based on a cultural/bioconservative ideology be different from an approach based on one of the traditions of liberalism? The most obvious difference is that a technocratic liberal approach would foreground a scientific study of natural systems. Like the romantic and emancipatory traditions of liberalism, it would also reinforce the cultural assumptions about a human-centered world and the progressive nature of change that have been the main supports of the Western form of hubris. The romantic and emancipatory liberals would continue to frame ecological literacy (if they were to consider it at all) in terms of individual freedom, and a form of critical reflection that is based on the assumptions that all traditions must be reconstituted to fit the subjective judgment of the individual. In effect, ecological literacy would either become part of the instrumental knowledge that leads to the technological problem solving and management or to the anomie type of individual who has proved so susceptible to the forces promoting consumerism.

Educational approaches to ecological literacy that are based on the assumptions of cultural/bioconservatism would be profoundly different. In addition to learning about the characteristics of natural systems, and the consequences of various forms of human intervention, ecological literacy would also have a more explicit cultural emphasis. That is, it would involve giving more attention to the adverse consequences for the environment of the cultural patterns taught in the non-science areas of the curriculum. Literature, history, and the arts, as well as all other areas of the curriculum, at both the public school and university level, represent a form of environmental education. That most subject areas continue to teach a destructive form of environmental education (or an attitude of indifference) can be seen in how little the non-science areas of the curriculum have changed in the face of the constant stream of media coverage about our deepening environmental problems. An examination of the curriculum from the elementary grades through graduate-level courses will reveal that the decline of forest cover, topsoil, fisheries (and now zooplankton), and the disappearance of both species and cultural languages that encode local knowledge of ecosystems have not altered the anthropocentrism and other liberal biases about ideas of modernity and individualism that continue to be passed on to the next generation.

First, there is a need to find a more appropriate balance between the current high-status forms of knowledge that contribute to the experimental orientation of modern cultures and the forms of knowledge about human and environmental relationships that some cultures have accumulated over hundreds or thousands of years.

A second area of educational reform involves shifting away from a liberal way of understanding creativity, intelligence, and the language-thought connection. Briefly, the current way of representing creativity and intelligence as an individual attribute, and language as a conduit through which we pass our ideas and information to others, contributes to a destructive form of environmental education. When students are socialized in this way of thinking, which is part of the Cartesian mind-set, they look at the world (both the cultural and natural) as separate from themselves. What they need to understand through all areas of the curriculum is that intelligence is primarily cultural in nature—in its embodied, taken for granted, and intentional/reflective dimensions. They need to understand that achieving meaningful and mutually supportive forms of community life that can exist within the limits of the environment is the ultimate test of cultural intelligence.

Furthermore, students need to understand how the coding and reproductive characteristics of language lead to "language thinking them" as they think within the language, to paraphrase Martin Heidegger's insight into the constitutive role of language. That is, they need to understand how the root metaphors and processes of analogic thinking of the past frame their current ways of thinking and acting. This would enable students to recognize how they are connected to the symbolic ecology of the culture and why past forms of intelligence may not be appropriate to understanding current relationships. When framed in terms of the need to conserve cultural patterns that are ecologically sustainable, this understanding of the culture-language-thought connection radically changes how we think about the content of every area of the curriculum.

Acknowledgment

An earlier version of this paper was presented as part of the Symposium on Natural Resource Issues and Education, sponsored by Utah State University.

References

Leopold, A. (1966). *Sand County almanac.* San Francisco: Sierra Club/Ballantine Books.

Orr, D. W. (1992). *Ecological literacy: Education and the transition to a postmodern world.* Albany: State University of New York Press.

By C. A. BOWERS

*Excerpt for Part I from: Bowers, C. A. (1996). The cultural dimensions of ecological literacy. The Journal of Environmental Education, 27(2), 5–10.

2   Co/inspiriting Ecological Conversations with Chet A. Bowers (1935–2017) and Ted T. Aoki (1919–2012)

Jennifer S. Thom

In the Middle

To situate the feature excerpt from Bowers' 1996 article *The Cultural Dimensions of Ecological Literacy* within the scope of his theoretical work is no trivial task. 1996 marks the midpoint of the scholar's 44-year focus on the crisis. Systemically speaking, many questions emerge: How do the ideas relate to Bowers' previous 22 years of theorizing? What ways do the concepts connect with those in his subsequent works, including his final book *Ideological, Cultural, and Linguistic Roots of Educational Reforms to Address the Ecological Crisis*, published exactly 22 years later in 2018? What relevance does Bowers' discourse hold for pressing issues today and for the future? What aspects of Bowers' theory call for further consideration to advance ecological approaches to curriculum discourse?

Coincidence or not, it was the summer of '96 when I started graduate studies at The University of British Columbia (UBC) in Vancouver, Canada. It was also while reading the first book for my class that Bowers' work caught my attention. The book was not one written by him, but rather, included one sentence which referred to him. In *The Web of Life*, author, theoretical physicist, and founding director of the Center for Ecoliteracy Fritjof Capra (1996) wrote: "C. A. Bowers has argued eloquently, language is metaphoric, conveying tacit understandings shared within a culture" (p. 70).

Fast forward two summers. Bowers accepts an invitation to teach a graduate course in the noted scholar series at UBC. The course focuses on his recently published book *The Culture of Denial* (1997), his critique of Western education, and reforms he views as necessary for public and postsecondary schools. Talk about an ecological approach to curriculum as (dis)course! Looking back, neither classmate Derek Rasmussen nor I could have predicted the years of conversations, mentoring, and friendship with Chet that followed.

Fast forward 18 more summers. Chet emails and asks me what I am working on. "Discourse, the everyday, ...life!" I reply. To this I add a list of authors, key ideas, and press "Send message." One of the authors is Ted Tetsuo Aoki and his notion of the live(d) curriculum. Aoki's (1993/2004) concept of the live(d) curriculum is "the more poetic, phenomenological and hermeneutic discourse in which life is embodied in the very stories and languages people speak and live" (p. 207). Chet writes back a few days later, explaining that while he did not know Aoki or his work,[1] he is very interested in the concepts and approach of this scholar's theorizing.

Chet also does not know Aoki was a UBC professor and former Director of the Centre for the Study of Curriculum and Instruction (CSCI)—the same Centre through which then newly appointed Director Karen Meyer had invited Chet to the university in 1998. The Centre is where Aoki taught as Emeritus Professor. Nearly eighty years old at the time, Aoki was still teaching and also mentoring Meyer (see Meyer, forthcoming). Together, Meyer and Aoki infused the curricular landscape (Aoki, 1993/2004) of the Centre with abiding humility, generosity, and brilliance.

Now just as then, while Aoki's live(d) theorizing of curriculum and pedagogy draws me in, it is *his* curriculum as live(d), which too is profound for me. Aoki's life as "nisei" (Japanese for first generation of children born in the new country to Japanese-born immigrants); being raised on Vancouver Island, Canada; a graduate of UBC; and his experiences during WWII, all hold strong resemblance and meaning to my family's history and thus to me as person of Japanese-and-Chinese heritage.

Of the handful of papers that Chet has read, he is most provoked by the ideas presented in Aoki's 1996 article *Spinning Inspirited Images in the Midst of Planned and Live(d) Curricula*. Intrigued, I hope we might explore them together. Unfortunately, yet understandably so, during the months leading up to his passing, Chet must attend to his final writings and see them through to publication.

Hindsight and the Need for Further Conversation

Fast forward another four summers. This chapter details what happened when I set out from the future—year 2020—to catch up with theoretical acquaintances of the past in Bowers' 1996 article. Bowers' question "Why has ecological literacy become so difficult for modern cultures to understand and carry out successfully?" (Bowers, 1996, p. 9) and the need he identifies regarding student understanding of the ways that "lead to 'language thinking them' as they think within language" (p. 9) immediately caught my attention. Both question and imperative present all the more urgent today given the environmental and cultural crises we face, as well, amidst COVID-19. While I wonder what conversations might be unfolding if Bowers was here today, the unexplored connections between his ideas and Aoki's persist. More curiously, what conversations might be emerging between Bowers and Aoki if they were both here now?

In this chapter, for the first time, Chet Bowers and Ted Aoki meet in dialogue. Direct quotations from their individual works come to life as a trio of conversations. Together Bowers and Aoki, two seemingly unlike scholars, revisit and co-theorize ideas they saw as vital to education. Emerging from their dialogue are concepts concerning the non-neutrality of language, curriculum as cultural discourse, and linguistic forms of colonization through print, data, and technology. It is these in-depth and far-reaching examinations which Bowers and Aoki each carried out in their theorizing that radically shifted curriculum thinking in the latter 20th century and still today in the 21st century. The performative nature of the conversations in this chapter emphasizes Aoki's signature turns, re-turns, and invocations to linger which deny closure and play off of Bowers' most certain, always impassioned, and at times, jarring responses. Here ideas surface and resurface as they are continuously taken up by the scholars and explored further in relation to the ecological crisis.

Other aspects of the chapter offer additional insight into Bowers' and Aoki's scholarship.

Chronologically connecting each of their published works gives rise to two timelines: Aoki's theorizing which spans four decades and Bowers' that extends across five. These chronologies enable further inquiry into the linear temporalities between Aoki's and Bowers' conceptualizations—that is, the parallels, overlaps, intersections, and divergences. In a complementary way, the simultaneous atemporal weaving of the presented texts illuminates the nonlinear manners in which Bowers and Aoki engage important ideas as curriculum theorists. For example, the conceptually common, complementary, relational, and co-emergent ways the scholars reveal and re-examine culturally taken-for-granted assumptions in curriculum.

This chapter thus endeavors to incite the theoretical and pedagogical sensibilities of these most influential scholars and *co/inspirit* (Aoki, 1996/2004) conversations about ecological approaches to curriculum discourse concerning our past, present, and future. The first two conversations serve as provocations to (re)consider concepts inherent in Bowers' 1996 question and of language as they encounter the ideas of Aoki's. From here, the exploration leads to an examination of emergent global challenges. Aoki and Bowers then invite us to join them in the third and concluding conversation to search for a place to begin anew; to understand thinking and think differently about ecological literacy, modernity, and the crisis. Arising from Aoki's queries related to *tensionality, a bridge which is not a bridge,* and *the space between* comes deeper meaning, newfound relevance, and transformative possibilities that resist enframing Bowers' theory of culture, language, and thought patterns. Conversation 1 opens with Bowers posing his question about ecological literacy.

Conversation 1: There's a Crisis and it's a Modern Ecological One[2]

BOWERS: Why has ecological literacy become so difficult for modern cultures to understand and carry out successfully (1996, p. 9)? When a problem is mentioned, such as pollution, the curriculum carefully isolates it by fostering a linear form of knowing whereby the student is encouraged to consider it without looking at the inter-relationships between pollution, technology, our form of economy, and our tendency to associate our standard of living with a view of society that tells them our basic assumptions are right, even though the media provides daily evidence to the contrary (1974, p. 61).

AOKI: It is analogous to the producer-consumer paradigm we have in business and industry. In this paradigm, specialists produce for those who consume…. curriculum experts produce programs for the consumers, the teachers and students (1974, p. 37–38). This crisis, manifested as a reflection of the contemporary image of advanced industrial society, shows as an internal crisis in curriculum (1983/2004, p. 114). Not to be able to see what is right is no error or deception; it is blindness (1987/1999/2004, p. 155). [In] our busy world of education, we are surrounded by layers of voices, some loud and some shrill, that claim to know… [with such a] cacophony of voices, certain voices bec[o] me silent and, hesitating to reveal themselves, conceal themselves (1987/1999/2004, p. 188). Pointedly, Heidegger (1977) says… what endangers [hu]man[s] is [their] inability to present other possibilities of revealing (1987/1999/2004, p. 153).

BOWERS: It's very important for teachers, parents, professors, theologians, whoever (2015c) to understand how the coding and reproductive characteristics of language—I use language as a verb (2015b)—lead to "language thinking them" as they think within the language, to paraphrase Martin Heidegger's insight into the constitutive role of language (Bowers, 1996, p. 9).

AOKI: "[O]ur cultural identities reflect the common historical references and shared cultural codes which provides, as one people, with stable, unchanging and continuous frames of reference and meaning…" (Hall as quoted in Aoki, 1996/2004, p. 315). Implied within a "perspective" are root metaphors, deep-seated human interests, assumptions… worldview, and knowledge… (1986/2004, p. 145). We live by metaphors. But most of the time we take our root metaphors for granted without realizing the assumptions we unconsciously hold. If we want to come to know the assumptions we make about humanity and world, we need to… make sense of our world by uncovering and thus discovering the root metaphor(s) to which we unconsciously subscribe (1979/2004, p. 346).

BOWERS: —to understand how the root metaphors and processes of analogic thinking of the past frame... current ways of thinking and acting (1996, p. 9).

AOKI: [C]urriculum scholars have opened themselves to the realm of language, ...discourse and narratives to understand their own field. Within this curricular turn, language is understood not so much as a disembodied tool of communication (1992/2004, p. 264)—

BOWERS: —The conduit view of language ...hides the basic reality that words have a history—as well as how they carry forward earlier culturally specific ways of thinking, misconceptions, prejudices, and silences (2013, p. 52).

AOKI: —[rather] language is understood in an embodied way—a way that allows us to say, "We are the language we speak" or "Language is the house of Being" (1992/2004, p. 264). What I see and how I see is because of who I am. I am what I see. I am how I see (1979/2004, p. 348).

BOWERS: [Whereas] the conduit view of languaging processes hides how words, as metaphors, encode earlier and culturally specific analogs that contribute to the linguistic colonization of the present by the past, and to the colonization of other cultures (2018, p. 47).

AOKI: Language is no mere communication tool; the very "languaging" participates in creating effects (2000/2004, p. 324). I recall... reading lessons with my Grade 1 students using a basal primer, *We Think and Do....* Naively... I was a blind reader, unable to read thoughts and ideas already inscribed in the text of *We Think and Do...* What a teacher of reading! Humiliating! Some decades later, ... a little wiser, maybe a little more humble... I can say that We Think and Do, as a version of "I think; therefore, I am," is historically grounded in the Age of Enlightenment, illuminating the shape and texture of the Western epoch we have come to know as Modernism. And, of course, we know how Western Modernism flourished as the disciplines of science and technology, which today hold a privileged position (1993/2004, p. 293).

BOWERS: [P]rint and data... are problematic; keep in mind, the Janus nature—they're positive, they're very important, they're indispensable. So, it's not either-or.... But the point is, what we need to challenge is this notion that what is *in* print is objective; it's factual.... What print and data cannot do for us is provide us a knowledge of context; context is your taken for granted world... memories... relationships, what is being negotiated in those relationships with others.... Encounter the printed word on a page... printed word on a computer screen.... What is being reinforced is a *conduit view of language...* [T]his is absolutely *essential* for maintaining that there is such a thing as objective knowledge; that there is factual information; that human beings are not complicit in the construction of that knowledge (2015a). What is lost... is that all language is metaphorical (emphasis added, 2015a).

AOKI: In this, it is not computer technology that is dangerous; it is the *essence* of computer technology that is dangerous (emphasis added, 1987/1999/2004, p. 153).

BOWERS: Increasingly, evolution is becoming a root metaphor and people are coming up with some really really frightening ideas—such as the computer science people who are highly educated in a narrow area but who are totally lacking an understanding of the cultures in which their technologies are introduced. But they see the... marginalization of jobs, the forms of knowledge that can't be encoded in print or data... as part of an evolutionary process... the post-biological phase of evolution... progress... [A]ll the books I've read on computer scientists talking about how the world is going to be changed and improved and globalized—they never mention what is being lost (2015a). They never mention what traditions we need to carry forward, what traditions we should conserve (2015b). But I think your ecological intelligence tells you you've lost privacy, you've lost security (2015a) [as well as] [t]raditions that have been overturned by both the digital revolution and the industrial revolution, where everything has to be monetized and contribute to the growth of markets (2015b).

AOKI: Who should be thinking, me or the computer?... [T]he predicament of humans who, humbled by the very machines we created are giving way most humbly to the machine's intelligence, artificial though it may be (1993/2004, p. 291).

BOWERS: This [point alone] would enable students to recognize how they are connected to the symbolic ecology of the culture and why past forms of intelligence may not be appropriate to understanding current relationships. When framed in terms of the need to conserve cultural patterns that are ecologically sustainable, this understanding of the culture-language-thought connection radically changes how we think about... every area of the curriculum (1996, p. 9).

AOKI: [T]he form of thinking that holds sway in most quarters wherein educators and non-educators dwell—... what we call rational thinking, logical thinking, or critical thinking, although in part it is (1992/2004, p. 196).

BOWERS: Enlightenment thinkers... argue what we need is a secular view of the world...rational thought... critical thought... and science (2015b).

AOKI: Thinking typically understood in our Western tradition has a seductively intellectual ring to it.... "Thinking" so understood is so familiar to us that when we say "thinking," we can think no other thought about thinking but that. In fact, we tacitly subscribe to this understanding of thinking such that we forget that we have been seduced into having a love affair with such an understanding. And in the blindness that usually accompanies such affairs, we fail to see other possibilities of understanding "thinking" (1992/2004, p. 196).

BOWERS: [I]n an era today where we're seeing carbon dioxide contributing to the acidification of the oceans where they're predicting the pH factor will drop to 7.8 by the end of this century, which means most of the coral will be gone *(pauses)*, most of the species will be gone in the water *(pauses again)*. We're really like that frog in the pot of water who isn't aware that the water is heating up. And our taken for granted beliefs are very much like that frog. We're not aware that... many of the misconceptions passed on from the past and in public schools and through our universities are based on a number of misconceptions and hubris... on a Western form of consciousness... [that] continues today...the priority of abstract thinking over... ecological intelligence...to ignore... those non-monetized, intergenerational, face-to-face, low toxic, low carbon footprint kinds of activities; what I'm calling the cultural commons (rearranged, 2015a). I[t]... make[s] the case for teachers giving more attention to the ethnographies of everyday experience rather than to passing forward the textbook misconceptions that most faculty still [rely] upon (personal communication, n.d.).

AOKI: [W]hat seems urgent for us at this time in understanding what teaching more truly is, [is] to... reorient ourselves so that we overcome mere correctness so that we can see and hear our doings... so we can see and hear who we *are* (1992/2004, p. 197).

BOWERS: The West is engaged in a very powerful process of colonization... we need to look at the process of linguistic colonization of our present by our past. And that is a tough nut to crack because like the frog, we're all very comfortable with our vocabulary—and damn it, we just don't want to take on the challenge of learning something that we haven't thought about before... that's the crisis we face (Bowers, 2015a).

The Significance of Ecological Literacy

Striking is the coherence between Bowers' and Aoki's ideas of culture-language-thought patterns. Even more remarkable is how neither the scholars nor their ideas met until Bowers read Aoki's articles in 2016—despite their shared theoretical interests overlapping four decades and both physically being in the same places at the same time—e.g., the University of Oregon, UBC, and the 2001 AERA Annual Conference—aptly themed "What We Know and How We Know It"!

In the conversation, Bowers (1996) introduces ecological literacy as he poses his question. Ecological literacy also appears in the title of the article and 13 more times throughout. However, because he does not define ecological literacy, its significance prior to 1996, during this period, and afterwards is unclear.

To be clear, ecological literacy was not just a term Bowers used but the thesis of 44 years of scholarship. By 1996 he had conceptualized ecological literacy for more than 20 years and continued to theorize it for more than 20 years afterwards. This was Bowers' exclusive focus to which he tirelessly devoted his self, time, and energy as curriculum scholar, teacher, mentor, colleague, friend, and human. His publications produced incisive and far-reaching ideas to understand, examine, and enact ecological literacy including what it means to be ecologically—*conscious* (Bowers, 2008), *sustainable* (Bowers, 1995), *conservative* (Bowers, 2003), *intelligent* (Bowers, 2011), *responsive* (Bowers & Flinders, 1990, Bowers & Flinders, 1991), *affirming* (Bowers, 2006), *post-liberal* (Bowers, 1987), and *just* (Bowers, 2001). In truth, ecological literacy was Bowers' curriculum vitae—his life's work.

Cultural Literacy

Bowers' attention to all things ecological goes back to his university years where Frederic Lilge, author of *The Abuse of Learning* (1948) and Bowers' doctoral supervisor, encouraged him to explore the relationship between language and realities. Bowers was also greatly influenced by the work of scientist, philosopher, and ecological conservationist Aldo Leopold (1949); *The Limits to Growth* authors Donella Meadows, Dennis Meadows, Jørgen Randers, and William Behrens (1972); and biologist Rachel Carson (1962). After earning his PhD, "it all came together" (Bowers, personal communication, n.d.) and Bowers self-published *Cultural Literacy for Freedom* (1974). Integrating his ideas with the ideas of these thinkers, Bowers revealed how education, culture, and the natural environment were inextricably entwined.

For Bowers, culture was not the passing down and carrying forth of reified traditions from generation to generation. Rather, cultures are continuously (re)created through the emergent and recursive communication of its members. Cultural literacy involves more than reading, writing, and speaking (Bowers, 1984). It extends beyond geographical and historical facts used to comprehend current events (Bowers, 1993). It requires understanding how continuity and change happen in the most minute (e.g., a word) to the most encompassing manners (e.g., climate change); how cultures affect one another; and how human cultures impact the more-than-human world (Abram, 1996). This cultural literacy is what Bowers saw as eventuating freedom. Freedom to comprehend life as "relationships, continuities, disjunctions, and trade-offs"

(Bowers, 1984, p. 2). Freedom to nurture the diversity of cultural and natural ecosystems. And freedom to make culturally and environmentally responsive change. Bowers contended this to be education's most important contribution and why he so vehemently urged educators to attend to these dimensions of culture.

From Cultural Literacy to Ecological Literacy

Bowers' 1974 book was widely read by educators as was his numerous subsequent publications, and to a certain extent, cultural issues were taken up. Despite his observation that "the environmental themes were totally ignored" (Bowers, personal communication, n.d.), this was not what concerned him most. Rather, it was the modern-day hubris which kept educators from recognizing their role in the crisis and concealed opportunities to educate for an ecologically just world. Bowers responded by exposing "three distinctive educational expressions of liberalism" (Bowers, 1996, p. 6)—romantic, technocratic, and emancipatory. These, he argued, over the decades and even posthumously, were the ideological and cultural roots of the crisis (e.g., Bowers, 1977, 1985, 2002, 2017, 2018).

Bowers' shift from cultural literacy to ecological literacy was not about trading in cultural issues for environmental ones but putting his theory into practice—enacting "words have a history" (Bowers, 1996, p. 6) and words "can be given new meanings when the choice of analogs is informed by other cultural ways of knowing and a knowledge of current environmental changes" (Bowers, 2013, p. 52). Bowers reconnected Greek meanings of "oikos" with "eco." "Household" reinforced individuals as members of collectives such as families, communities, cultures, nature, and earth (e.g., Bowers, 1995) whereas as "the operations and management of the family household" (Bowers, 2005, p. 158) emphasized those relationships that sustain such collectives. Revitalizing oikos to reconceptualize ecology brought the culture-environment connection front and center, allowing him to address cultural issues *with* environmental ones. Throughout Bowers' theorizing is his conception of sustainability as intelligence or "the larger patterns of interaction and interdependence that we find in community, culture, and the way in which cultures are embedded in natural systems" (2005, p. 158). This meaning along with those developed in the two literacies proved pivotal to Bowers' later explication of ecojustice, his reconceptualization of the cultural and environmental commons, and championing issues imperative for their revitalization.

Linguistic Dimensions of Ecological Literacy: Culture-Language-Thought Patterns

In the conversation when Aoki (1987/1999/2004) states, "Not to be able to see what is right is no error or deception; it is blindness" (p. 155) and Bowers (2015a) refers to the unaware frog, these are not mere figures of speech or fear tactics. Rather, they signal the crisis as inherently culture-language-thought patterns wherein resolution requires making the invisible visible (Merleau-Ponty & Lefort, 1968), the familiar strange (Nietzsche, 1998), and the strange familiar (Deleuze, 2000). Each scholar draws on different works of Heidegger's, underscoring the crisis as fundamentally phenomenological. For Aoki (1987/1999/2004), his focus concerns things as they (dis)appear and the inability to reveal what has been covered, particularly within the context of understanding the computer and its application as technology (Heidegger, 1977). For Bowers (1996), it is Heidegger's (1962) notion of "'*language thinking them*' as they think within the language" (emphasis added, p. 9). Here he re-enacts the Greek meanings of oikos as he infuses the ideas of house and the maintaining of its relations into a broader critique of socio-political *eco*nomy.[3]

Aoki and Bowers' critical dialogue on cultural codes and how language is never objective or neutral elucidates the relational and inconspicuous ways cultural languages language. Along with this, as human communication depends on specific metaphors for shared meaning, these communicative structures therefore must be understood as ecologies, oikos, every culture's "house of being" (Aoki, 1992/2004, p. 264). Bowers' theory on language is essential to revealing the complexity of culture-language-thought patterns we do not see (Aoki, 1987/1999/2004), are unaware of (Bowers, 2015a), or take for granted—precisely what Capra meant by "C. A. Bowers has argued eloquently, language is metaphoric, conveying *tacit* understandings shared within a culture" (emphasis added, 1996, p. 70).

This work of Bowers dates back to the 1970s when he began his examination of the metaphoric nature of language. Included was sociologist Richard H. Brown's (1977) concepts of root, analogic, and iconic metaphors related to cultural (re)production. Root metaphors emerge from particular events (e.g., inventions, wars, COVID-19) and mythopoetic narratives (e.g., creation stories, Big Bang theory, survival of the fittest). Over time and undergoing metaphorization, the meanings from such events take "root" in the culture's communicative processes, giving rise to analogic (e.g., earth as mother, brain as computer, humans as land) and iconic metaphors (e.g., relationality, genetic engineering, gig economy), all which define the thinking, discourse, and conscience of the collective. Not only did Bowers (e.g., 1979, 1984, 2002, 2012, 2018; Bowers & Flinders, 1990, Bowers & Flinders, 1991) reveal how metaphors promoted in education were identical to those in dominant society—e.g., mechanism, scientism, economism, individualism, anthropocentrism, and colonialism; in radical ways, he shed light on how these metaphors, many which are centuries and even millennial years old, separate humans from rest of the world and privilege Western cultures. These metaphors and the ideas, thinking, and actions engendered by them are what Bowers conceived as the linguistic roots of the crisis which educators, the curriculum, and system at large (un)consciously perpetuate.

Bowers' theory on culture-language-thought patterns is arguably his most profound gift to education and indeed to curriculum studies. It allows for examination of the ideological, cultural, and linguistic manners which discourse generally and particularly "thinks us as we think within it" (Bowers, 1996; Heidegger, 1962). His theory compels deep collective inquiry into "the assumptions we make of humanity and the world" (Aoki, 1979/2004, p. 346)—what it means to be human and alive; to be earth (Heidegger, 1971); and what might enable "dwelling alright within it" (Aoki, 1986/1991/2004, p. 163). Carrying these considerations forward, I explore Bowers' ecological literacy as it relates to today's emerging global challenges.

A Curriculum-as-Plan(ned) for Imagining Canada's Future

"The curriculum-as-plan[ned] is ...imbued with the planners' orientations to the world, ...their own interests and assumptions... usually implicit in the text of the curriculum-as-plan, frame a set of curriculum statements" (emphasis added, Aoki, 1993/2004, p. 202).

In 2018, Federal Government organization Policy Horizons Canada (PHC) published *The Next Generation of Emerging Global Challenges: A Horizons 2030 Perspective on Research Opportunities*. As part of the Social Sciences and Humanities Research Council's initiative for *Imagining Canada's Future*, the objectives for this curriculum-as-plan(ned) (Aoki, 1986/1991/2004) are set in front of the following backdrop: "Climate change is a familiar global challenge. While that problem is far from solved, we were focused on what else lies over the horizon" (PHC, 2018, p. 1). Six of the 12 initial challenges include: *humanity+, the evolving bio age, truth under fire in a post-fact world, the erosion of culture and history, the pervasive contamination of the "natural," living within the carrying capacity of planet earth*, and *working in the digital economy*. Each challenge (in)directly leads to one or more of the accumulative challenges: *inhabiting challenging environments, the emerging asocial society, envisioning governance systems that work*, and *the arts transformed*.

The document as primer, much like the basal reader referred to by Aoki (1993/2004), not only projects a global scan but also directs current and future foci as well as defines funding opportunities for researchers. Thus, it is no mere communication tool (Aoki, 2000/2004) or conduit of text (Bowers, 2015a). Reading and critically interpreting this curriculum-as-plan(ned) requires understanding the ideas as well as the culture-language-thought patterns inscribed in it and prescribed by it. Consider the challenges of *humanity+* and *the bio age*. Framed within a Western view of progress (Bowers, 2015a), science and technology "unlocks human potential and redefines what we mean by humanity" where "[i]ncreasing understanding of complex living systems enables inventions that borrow from designs found in nature, innovations that blur the boundaries between micromachines and cells, and the redesign and synthesis of living organisms" (PHC, 2018, p. v). Moreover, "[t]hese contribute to a possible future with more ecologically benign technologies, living machines and sensors, and deliberately designed human characteristics" raising concerns of "regulatory challenges and new vulnerabilities such as body system sabotage hacks, as people embed technologies, along with entirely new dimensions of inequality and biased access" (PHC, 2018, p. 7).

What once read as science fiction now reads prescriptively as the curriculum-as-plan(ned) for social sciences and humanities research in Canada. And while cultural industrialization (un)consciously plays forward as the digital revolution (Bowers, 2016), there is another danger; a languaging loophole of learned helplessness[4] (Seligman, 1975)—that we are past the point of no return regarding saving the planet, cultural erosion, and the engineering of our species. The advancement of the West's colonization (Bowers, 2015a) holds as a viable future, one that increasingly requires digital worlds, virtual communities, new "human tribes… courtesy of human augmentation and bio-design," and "new tribes of robots and their digital ecosystems" (PHC, 2018, p. vi).

Turning to Bowers and Aoki, they take up these issues as Bowers' original question is revisited in the second conversation. This time, however, it becomes clear that the answer will be anything but straightforward.

Conversation 2: "The Uncannily Correct but Not yet True"[5]

BOWERS: There is an increasing number of people… who are aware there is an environmental problem (2015b).

AOKI: We are less naïve today (1992/2004, p. 189).

BOWERS: [B]ut they are unable to address what we can do that's different from the current patterns that are pushing us beyond the environmental limits (2015b).

AOKI: [W]e see about us efforts… akin to a technological understanding of teaching whose logical outcome is…robotization of teaching: schools in the image of Japanese automobile factories—heaven forbid! (1992/2004, p. 189).

BOWERS: Why has ecological literacy become so difficult for modern cultures to understand and carry out successfully (1996, p. 9)?

AOKI: Many… have approached the question…from their own favoured perspectives…. that provide some understanding of human doings…within the grasp of reasoned control. They present indeed, a seductively scholarly and intellectual quality and legitimacy that makes the understanding of teaching uncannily correct (1992/2004, p. 189). For Lacan, the discourse of the master doctor and the patient is inadequate; instead, he opts for the to- and-fro discourse of teaching/learning. For him, listening to "what" is being said requires listening to "where" the "what" is being said. Then, the "what" can be interpreted in terms of the "where" (2000/2004, p. 325).

BOWERS: [M]isunderstanding results from viewing education only as occurring in public schools or universities. To understand the complexity of the educational processes that contribute to various and often contradictory forms of environmental education, it is necessary to recognize that education, in the broadest sense, is synonymous with culture.... how the assumptions, values, technologies, and categories of thinking of a culture influence the way humans relate to the environment (1996, p, 5).

AOKI: [C]ould it be ... a product of a collision of things that refuse to be bound together neatly. They clang about, not fitting right. So understood, could it be that... the structure of a bind, a site of tension between this and that, a site of difference that speaks of two or more things at the same time? Could it be that what is at work is a situational paradox of sorts? Could it be that awareness of binds is a call to remind us that we, as humans, live in a divided way, in a realm of both this and that (1993/2004, p. 291)?

BOWERS: The ecological crisis is not really being understood in terms of how imminent it is—that is... the melting of glaciers, the dry-outs, the extreme weather conditions.... Now, what's the connection between language and the ecological crisis? The basic question is, to what extent do the words that encode earlier assumptions and ways of thinking continue to influence our current ways of thinking? And so, to get at that, one needs to understand the metaphorical nature of language.... And I've heard scientists give very good explanations about different phenomena but when it comes to telling us what are some of the solutions to this problem, they go silent. So, one of the solutions to this problem is to address how it is that language reproduces these earlier patterns of thinking; language that we take for granted as we are born into those language communities and they become the basis of our thinking (2015b).

AOKI: Could it be that with over 40 years of teaching I have become preoccupied with so many answers to the question... that I have forgotten to question my own understandings of the question itself? Could it be that in the years of questioning...I have come to an understanding not so much of what teaching is, but rather what teaching is not? Could it be that this sort of understanding—a negative understanding—is a stage of understanding that allows us to begin to see the uncannily correct but not yet true (1992/2004, p. 187)?

Questioning Understandings of the Question Itself

Different from the first conversation, two contrasting metalogues surface here. While Bowers lays out further aspects of ecological literacy, Aoki reflects on what teaching is. It is only after posing several further questions that Aoki realizes he has neglected to consider his own thinking about the original question. Could this be true for Bowers' question about ecological literacy? How does Aoki's insight and queries offer possibilities for understanding thinking—here, Bowers' more than 40 years of theorizing in relation to his 1996 question?

(Re)Thinking Bowers' Question as We (Re)Think Within It

It seems fitting to turn from examining discursive aspects that answer Bowers' question to now explore how the question thinks us as we think within it (Bowers, 1996; Heidegger, 1962). In other words, how does understanding Bowers' discourse in this manner deepen meaning for the question itself?

Bowers' question "Why has ecological literacy become so difficult for modern cultures to understand and carry out successfully?" reveals a problem in need of a solution. And over the years, as exemplified in the two conversations, he identified key challenges of modern society, provided reasons for them, and explicated how to overcome them. Yet how is it after Bowers' decades-long theorizing and 50 years of Earth Days that "we have made little progress on the most significant threats, including climate change[?]" (PHC, 2018, p. 53). In all seriousness, what becomes so difficult is persisting with the idea that modern society simply cannot recognize its destructive culture-language-thought-patterns or that because we are too comfortable in our modern ways, we do not want to learn something we have not thought about before (Bowers, 2015a). Might there be other discursive challenges contributing to this difficulty and could they include ones inherent in ecological literacy itself? If so, how do they impede responses to the crisis and in turn make addressing "the next generation of emerging global challenges" (PHC, 2018, p. i) that much more difficult?

Dichotomy, Enclosure, and Double Binds

Bowers' response to the crisis required ecological literacy be distinct from modernity. However, achieving this leaves the discourses almost if not wholly separate. Clearly, this was not Bowers' intention as he knew well, the limits of either-or thinking. Yet despite his efforts to locate "leverage points" (Bowers, 1997); rectify meanings of everyday words; revitalize the cultural and environmental commons, especially in Western cultures; and confront modernity head-on, in print—to challenge "seductively scholarly and intellectual" (Aoki, 1992/2004, p. 189) hubris which continuously legitimize and privilege print and data-based knowledge over all other forms of intelligence (Bowers, 2015a) ironically, the dichotomy continues. Perhaps even more pronounced today, it demonstrates the difficulty of understanding thinking and thinking differently.

With this, a kind of theoretical enclosure results from Bowers' explication and championing of the tenets used to frame ecological literacy (e.g., Bowers 1996) and later ecojustice (e.g., Bowers, 2001). What becomes so fundamentally distinct about the discourse is also what becomes so deeply difficult for educators as members of modern society to comprehend and reconcile, much less carry out successfully. Bowers' recognition of the difficulty is implied when he says that people are more aware of environmental (and cultural) problems yet they remain at a loss of what can be done differently to address them (2015b).

Similarly, while social, cultural, and environmental problems are identified in *The Next Generation of Global Challenges* (PHC, 2018), the tacit conceptual frames that portray and communicate them limit thinking deeply and as such, restrict opportunity for researchers to address them more fully. For example, regardless of global challenge, these prescriptive challenges reinforce each other in serving as means toward ends (see PHC, 2018, p. ii). Conceptually relating them as portrayed then precludes understanding the very question itself (Aoki, 1992/2004). Not attending to the assumptions or consequences of such forward-minded thinking effectively disregards recursive possibilities for research. Even deeper, unquestioningly framing research as forward-moving denies recognizing and attending to what "Bowers has argued eloquently" (Capra, 1996, p. 7); that is, how such thinking exemplifies "*tacit* understandings shared within a culture" (emphasis added, Capra, 1996, p. 70).

Enframing thinking as such gives rise to other communicative "languaging" (Aoki, 2000/2004, p. 324; Bowers, 2018, p. 47) that inherently "participates in creating effects" (Aoki, 2000/2004, p. 324) within the actual challenges themselves. Continuing not to pay attention to culture-language-thought patterns (Aoki, 1992/2004; Bowers, 2018) suggests they persist in determining how we perceive and understand the global challenges. These tacit patterns thus bypass systemic issues that connect one challenge to another and in doing so, foreclose viable solutions for addressing them all. Consequently, prospects for ecological thinking to inform modern conceptions and vice versa seem unlikely. It is more likely that theoretical enclosures as these further obscure implicit assumptions of this curriculum-as-plan(ned) (Aoki, 1993/2004), exacerbating the divide between the two discourses, and making the traversing of them less tenable.

Moreover, anthropologist and cyberneticist Gregory Bateson's (1972) double bind theory elucidates deeper hidden dynamics regarding the PHC (2018) curriculum. Simply put, a double bind is "a situation in which no matter what a person does, [t]he[y] 'can't win'" (Bateson, Jackson, Haley, & Weakland, 1956, p. 251). Applying Bateson's logic to interpreting the PHC (2018) text discloses a double bind that posits as the underlying assumption and starting point: "Climate change is a familiar global challenge. While that problem is far from solved, we were focused on what else lies over the horizon" (PHC, p. 1). Here, over-looking climate change as a familiar global challenge creates internal conflict within the curriculum-as-plan(ned)—namely, how it relates to the ecological crisis. Social sciences and humanities research as seen through Aoki's theorizing of curriculum-as-plan(ned) and informed by his concepts of curriculum-as-live(d) further illuminates the double bind. Such learned helplessness (Seligman, 1975) toward climate change, as imbued by the planners' orientations to the world (Aoki, 1993/2004), serves to frame, set, and negate climate change as a global challenge and its connection to the crisis.

There's a Crisis. It's a Modern and Ecological One

Today's silence and inaction concerning the crisis cannot entirely be a lack of awareness or willful resistance but suggests a double bind which creates discursive inability to reveal genuine possibilities for under-standing thinking and thinking differently. An initial examination of Bowers' 1996 question discloses some of what else might be "so difficult" between the two discourses. Emergent here is a crisis discursively mod-ern *and* ecological as a site of tension and consequently, an impasse. Using Aoki's words, the ideologies, mythopoetic narratives, and meta-phors of the discourses clang and collide against each other, speak "two or more things at the same time," and "refuse to be bound together neatly" (Aoki, 1993/2004, p. 291). Bowers' scholarship expounds eco-logical literacy in contrast to modernity. Yet theoretical limitations arise when what modernity and ecological literacy are or could be is con-strained by what the other is not, hence the double bind. What kinds of thinking are possible when these understandings, now even more important are understood as "uncannily correct but not yet true" (Aoki, 1992/2004, p. 153; Heidegger, 1977, p. 6)?

Conversation 3: In (Ano)the(r) Middle

Aoki (1993/2004) reminds us that "awareness of *binds* is a call... that we, as humans, live in a *divided* way, in a realm of both this and that" (emphasis added, p. 291). Imagining a third conversation, Aoki and Bowers invite us to join them wherein Aoki proposes modernity and ecological literacy be connected as "two worlds by a bridge" (Aoki, 1981/2004, p. 228). Here he explains this as not "[m]erely to describe and characterize physical bridges and their metaphorical extensions in transportation and communications" (Aoki, 1988/1991/2004, p. 438), but to explore bridges in Oriental gardens and contemplate their meanings. Such structures so architecturally designed and metaphorically intended draw attention to the physical and mortal as well as the mythopoetic and immortal landscape. Built not to be straight but to vary in height, curve, angle, and slope, these bridges thus allow different areas of the garden to be experienced from multiple perspectives.

Stepping onto this bridge shifts our gaze to how modernity and ecological literacy (can) relate in ways other than an impasse. The particularities of each discourse as well as their contextual and complicit situatedness take on new prominence. And although modernity and ecological literacy extend in different directions, Aoki's (1987/1999/2004) bridge which structures thinking as "is and yet to be" welcomes convergences of the two as well as how the discourses, in tandem, are integral to the larger curricular landscape (Aoki, 2000/2004).

Walking a little further, Aoki (1981/2004) shows how this bridge is "a bridge which is not a bridge" (p. 228). True bridges, he contends, are not "mere paths... mere routes..." (Aoki, 1999/2004, p. 438) to overcome obstacles but places which free us to dwell on the earth (Heidegger, 1971) where "on this bridge, we are in no hurry to cross over; in fact, such bridges lure us to linger" (Aoki, 1996/2004, p. 316). What is more, while the meaning for dwelling is connected to the Gothic *wunian* for peace and to the German *friede* for free (Heidegger, 1971), "hashi" which is Japanese for "bridge" is also the homonym for "edge" (Keane, 1996, p. 171). This bridge as place to linger and dwell is also an edge where ecological literacy *meets* modernity. Here lingering does not imply compromise, simple appreciation, or taking the best of both worlds. Instead, "dwelling alright within it" (Aoki, 1986/1991/2004, p. 163) means bringing forth new space that compels working creatively *with-and-in* discourses, making generative use of the tension. Aoki (1996/2004) says, "[s]uch spaces are *edgy* spaces, located at margins and boundaries, spaces of doubling, where 'this or that' becomes 'this and that,' ambiguously, ambivalently—difficult places but nonetheless spaces of generative possibilities" (emphasis added, p. 422). An edge, then, must not be seen as periphery but as "and," a middle, or more articulately, "the traversal of the shifting spaces" (Aoki, 1996/2004, p. 422) where modernity and ecological literacy as "this-and-that" can be conceptualized as radically different while also linguistically understood as whole.

Looking out from this bridge, the next generation of global challenges (PHC, 2018) have transformed. Now viewed as bridges, they disclose a curriculum landscape in which *inhabiting challenging environments, the emerging asocial society, envisioning governance systems that work, and the arts transformed* (PHC, 2018) are no longer ends that preclude. Recursively connected as bridges and landscape, all challenges linguistically re-emerge as modern *and* ecological spaces of generative possibility.

How can sciences, technologies, economies, intelligences, cultures, and natural ecosystems inseparably as "this-and-that," be(come) infinite sources of mutual support and co-generativity within finite limits of here-now-and-future-generations? How as seen from this bridge might we conceptualize *economies*—those qualitative, non-monetized, face-to-face, local, nonlinear, intergenerational, and cooperative as forming edges with digital ones? In what ways does lingering here open generative spaces for diversity that occasions interrogation of otherwise uncritical prior acceptance of *the emergence of asocial society*, the need for *inhabiting challenging environments*, and *the pervasive contamination of the "natural"*? (PHC, 2018) How does committing to "dwelling alright within it" (Aoki, 1986/1991/2004, p. 163)—such edges—create opportunity to re-metaphorize modernity *and* ecological literacy amidst reimagining these next global challenges? Further still, rather than foregone conclusions, what futures vibrant and alive can flow when discursive worlds are connected by "both bridges and non-bridges" (Aoki, 1996/2004, p. 318)? In Aoki's words "authentic dwelling, as Heidegger would say, [is] made possible by the way mortals are, on this earth beneath the sky, beings who belong together… [or] [w]hen Inazo Nitobe spoke of his wish to serve as a bridge, his meaning was surely more than a physical structure than connects two masses of land. He spoke of what a bridge means humanly" (Aoki, 1988/1991/2004, p. 438).

Looking on, Aoki asks, "Where are we" (Aoki, 1993, p. 298)?

"[T]he challenge of learning something we haven't thought about before" reminds Bowers (2015a).

"I wonder" says Aoki (1993/2004, p. 298), then pauses.

Recalling Deleuze (1987), he knows a curricular "landscape of multiplicity" (Aoki, 1992/2004, p. 277) "like rhizomean plants, shoot from here to there, and everywhere working through, nourished by the humus" (Aoki, 1996/2004 p. 413). Such a landscape "grows from the middle" (Aoki, 1993/2004, p. 207). Learning something we have not thought about before calls for thinking that resists stasis and concealment but also denies determinism and (en)closure.

Continuing, Aoki re-turns with an open question in which for us to linger (Aoki, 1996/2004) and dwell (Heidegger, 1971).

"I wonder… [p]ositioned in an 'and,' (Aoki, 1993/2004, p. 298) lingering in this space of *live[d]*[6] tensionality (emphasis added, Aoki, 1993/2004, p. 300), if *and* might be a place where we can think differently. Can it be a place where human's *[E]go* can become decentered, can become dissolved a bit? [H]ow shall we begin to think anew?" (Emphasis added, rearranged, Aoki, 1993/2004, p. 298.)

Notes

1. This was most likely due to their connections with different academic communities and that many of Aoki's papers, in the past, were either unpublished or not readily available in the United States.
2. Bowers (2015b).
3. For example, see Bowers (1996), p. 38–39 for discussion on "The Modernizing Orientation of Universities," featuring the work of Karl Polanyi.
4. Seligman (1975) posits, "According to learned helplessness theory, people exposed to uncontrollable events learn that their responses and outcomes are independent of each other. This learning can lead to an expectation that responses will be futile and can generalize to new situations to interfere with future learning" (p. 681).
5. Aoki (1992/2004, p. 153).
6. While "live(d)" does not appear in this published work of Aoki's, it is consistent with his later writings. For example, see Aoki (1996/2004).

References

Abram, D. (1996). *The spell of the sensuous: Perception and language in a more-than-human world.* New York, NY: Vintage Books.

Aoki, T. T. (1974). Pin-pointing issues in curriculum decision-making. *In curriculum decision making in Alberta: A Janus look* (pp. 24–42). Edmonton, AB: Alberta Department of Education.

Aoki, T. T. (1979/2004). Reflections of a Japanese Canadian teacher experiencing ethnicity. In W. F. Pinar & R. L. Irwin (Eds), *Curriculum in a new key: The collected works of Ted T. Aoki* (pp. 333–348). Mahwah, NJ: Lawrence Erlbaum Associates, Inc.

Aoki, T. T. (1981/2004). Toward understanding curriculum: Talk through reciprocity of perspectives. In W. F. Pinar & R. L. Irwin (Eds), *Curriculum in a new key: The collected works of Ted T. Aoki* (pp. 219–228). Mahwah, NJ: Lawrence Erlbaum Associates, Inc.

Aoki, T. T. (1983/2004). Curriculum implementation as instrumental action and as situational praxis. In W. F. Pinar & R. L. Irwin (Eds), *Curriculum in a new key: The collected works of Ted T. Aoki* (pp. 111–123). Mahwah, NJ: Lawrence Erlbaum Associates, Inc.

Aoki, T. T. (1986/1991/2004). Teaching as indwelling between two curriculum worlds. In W. F. Pinar & R. L. Irwin (Eds), *Curriculum in a new key: The collected works of Ted T. Aoki* (pp. 159–165). Mahwah, NJ: Lawrence Erlbaum Associates, Inc.

Aoki, T. T. (1986/1996/2004). Interests, knowledge and evaluation: Alternative approaches to curriculum evaluation. In W. F. Pinar & R. L. Irwin (Eds), *Curriculum in a new key: The collected works of Ted T. Aoki* (pp. 137–150). Mahwah, NJ: Lawrence Erlbaum Associates, Inc.

Aoki, T. T. (1987/1999/2004). Toward understanding computer application. In W. F. Pinar & R. L. Irwin (Eds), *Curriculum in a new key: The collected works of Ted T. Aoki* (pp. 151–158). Mahwah, NJ: Lawrence Erlbaum Associates, Inc.

Aoki, T. T. (1988/1991/2004). Bridges that rim the Pacific. In W. F. Pinar & R. L. Irwin (Eds), *Curriculum in a new key: The collected works of Ted T. Aoki* (pp. 437–439). Mahwah, NJ: Lawrence Erlbaum Associates, Inc.

Aoki, T. T. (1992/2004). Layered voices in teaching: The uncannily correct and the elusively true. In W. F. Pinar & R. L. Irwin (Eds), *Curriculum in a new key: The collected works of Ted T. Aoki* (pp. 187–198). Mahwah, NJ: Lawrence Erlbaum Associates, Inc.

Aoki, T. T. (1993/2004). Legitimating lived curriculum: Toward a curricular landscape of multiplicity. In W. F. Pinar & R. L. Irwin (Eds), *Curriculum in a new key: The collected works of Ted T. Aoki* (pp. 199–215). Mahwah, NJ: Lawrence Erlbaum Associates, Inc.

Aoki, T. T. (1996/2004). Spinning inspirited images. In W. F. Pinar & R. L. Irwin (Eds), *Curriculum in a new key: The collected works of Ted T. Aoki* (pp. 413–423). Mahwah, NJ: Lawrence Erlbaum Associates, Inc.

Aoki, T. T. (2000/2004). Language, culture, and curriculum.... In W. F. Pinar & R. L. Irwin (Eds), *Curriculum in a new key: The collected works of Ted T. Aoki* (pp. 321–329). Mahwah, NJ: Lawrence Erlbaum Associates, Inc.

Aoki, T. T. (2000/2003/2004). Locating living pedagogy in teacher "research": Five metonymic moments. In W. F. Pinar & R. L. Irwin (Eds), *Curriculum in a new key: The collected works of Ted T. Aoki* (pp. 425–432). Mahwah, NJ: Lawrence Erlbaum Associates, Inc.

Bateson, G. (1972). *Steps to an ecology of mind.* Chicago, IL: The University of Chicago Press.

Bateson, G., Jackson, D. D., Haley, J., & Weakland, J. (1956). Toward a theory of schizophrenia. *Behavioural Science, 1*(4), 251–264.

Bowers, C. A. (1974). *Cultural literacy for freedom: An existential perspective on teaching, curriculum, and school policy.* Eugene, OR: Elan Northwest Publishers.

Bowers, C. A. (1977). Cultural literacy in developed countries. *Prospects, 7*(3), 323–335.

Bowers, C. A. (1979). The ideological-historical context of an educational metaphor. *Theory Into Practice, 18*(5), 316–322.

Bowers, C. A. (1984). *The promise of theory: Education and the politics of cultural change.* New York, NY: Longman.

Bowers, C. A. (1985). Culture against itself: Nihilism as an element in recent educational thought. *American Journal of Education, 93*(4), 465–490.

Bowers, C. A. (1987). *Elements of a post-liberal theory of education.* New York, NY: Teachers College Press.

Bowers, C. A. (1996). The cultural dimensions of ecological literacy. *The Journal of Environmental Education, 27*(2), 5–10.

Bowers, C. A., & Flinders, D. J. (1990). *Responsive teaching: An ecological approach to classroom patterns of language, culture.* New York, NY: Teachers College Press.

Bowers, C. A., & Flinders, D. J. (1991). *Culturally responsive teaching and supervision: A handbook for staff development.* New York, NY: Teachers College Press.

Bowers, C. A. (1993). *Education, cultural myths, and the ecological crisis: Toward deep changes.* Albany, NY: State University of New York Press.

Bowers, C. A. (1995). *Educating for an ecologically sustainable culture: Rethinking moral education, creativity, intelligence, and other modern orthodoxies.* Albany, NY: State University of New York Press.

Bowers, C. A. (1997). *The culture of denial: Why the environmental movement needs a strategy for reforming universities and public schools.* Albany, NY: State University of New York Press.

Bowers, C. A. (2001). *Educating for eco-justice and community.* Athens, GA: University of Georgia Press.

Bowers, C. A. (2002). Toward an eco-justice pedagogy. *Environmental Education Research, 8*(1), 21–34. https://doi: 10.1080/13504620120109628.

Bowers, C. A. (2003). *Mindful conservatism: Rethinking the ideological and educational basis of an ecologically sustainable future.* Washington, DC: Rowman & Littlefield Publishers.

Bowers, C. A. (2005). Afterword. In C. A. Bowers & F. Apfell-Marglin (Eds.), *Re-thinking Freire: Globalization and the environmental crisis (pp. 151-192).* Mahwah, NJ: Routledge.

Bowers, C. A. (2006). *Revitalizing the commons: Cultural and educational sites of resistance and affirmation.* Lanham, MD: Lexington Books.

Bowers, C. A. (2008). *Toward a post-industrial consciousness: Understanding the linguistic basis of ecologically sustainable educational reforms.* Eugene, OR: Eco-Justice Press LLC.

Bowers, C. A. (2011). *Perspectives on the ideas of Gregory Bateson, ecological intelligence, and educational reforms.* Eugene, OR: Eco-Justice Press LLC.

Bowers, C. A. (2012). Questioning the idea of the individual as an autonomous moral agent. *Journal of Moral Education, 41*(3), 301–310. https://doi: 10.1080/03057240.2012.691626.

Bowers, C. A. (2013). *In the grip of the past: Educational reforms that address what should be changed and what should be conserved.* Eugene, OR: Eco-Justice Press LLC.

Bowers, C. A. (2015a, March 30). How the worlds of data destroy ecological intelligence. *Youtube.* https://www.youtube.com/watch?v=PgM-kGB1foI

Bowers, C. A. (2015b, September 14). Part 1: Linguistic roots of the ecological crisis. *Youtube.* https://www.youtube.com/watch?v=O0YSPtPnNio

Bowers, C. A. (2015c, September 25). Part 2: Linguistic roots of the ecological crisis. *Youtube.* https://www.youtube.com/watch?v=ej6EG9fm70E

Bowers, C. A. (2016). *A critical examination of STEM: Issues and challenges.* Mahwah, NJ: Routledge.

Bowers, C. A. (2017). Educational reforms for survival. *Tikkun, 32*(4), 25–32.

Bowers, C. A. (2018). *Ideological, cultural, and linguistic roots of educational reforms to address the ecological crisis: The selected works of CA (Chet) Bowers.* Mahwah, NJ: Routledge.

Brown, R. H. (1977). *A poetic for sociology: Toward a logic of discovery for the human sciences.* Cambridge, UK: Cambridge University Press.

Capra, F. (1996). *The web of life: A new scientific understanding of living systems.* New York, NY: Anchor Books.

Carson, R. (1962). *Silent spring.* Boston, MA: Houghton Mifflin.

Deleuze, G. (1987). *Bergsonism* (T. Conley, Trans.) Brooklyn, NY: Zone Books.

Deleuze, G. (2000) *Proust and signs: The complete text* (R. Howard, Trans.). Oxford, UK: Athlone Press.

Meadows, D. H., Meadows, D. L., Randers, J., & Behrens, W. W. III (1972). *The limits to growth: A report for the club of Rome's project on the predicament of mankind.* New York, NY: Universe Books.

Heidegger, M. (1962). *Being and time* (J. Macquarrie & E. Robinson, Trans.). New York, NY: Harper Collins Publishers.

Heidegger, M. (1971). Building dwelling thinking. In *Poetry, language, thought* (A. Hofstadter, Trans.) (pp. 143–159). New York, NY: Harper Collins Publishers.

Heidegger, M. (1977). The question concerning technology. In *The question concerning technology and other essays* (W. Lovitt, Trans.) (pp. 3–35). New York, NY: Garland Publishing.

Keane, M. P. (1996). *Japanese garden design.* North Clarendon, VT: Tuttle Publishing.

Leopold, A. (1949). *A Sand County almanac.* Oxford, UK: Oxford University Press.

Lilge, F. (1948). *The abuse of learning: The failure of the German University.* New York, NY: Macmillan Company.

Merleau-Ponty, M., & Lefort, C. (1968). *The visible and invisible* (A. Lingis, Trans.). Evanston, IL: Northwestern University Press.

Meyer, K. (forthcoming). Interlude: Letters from Ted. In N. Y. S. Lee, J. Ursino, & L. Wong (Eds.) *Lingering with the works of Ted T. Aoki: Historical and contemporary significance for curriculum research and practice.* Mahwah, NJ: Routledge.

Nietzsche, F. (1998). *On the genealogy of morality* (M. Clark, & A. J. Swensen, Trans.). Indianapolis, IN: Hackett Publishing Co.

Policy Horizons Canada (PHC). (2018). *The next generation of emerging global challenges: A horizons 2030 perspective on research opportunities.* Ottawa, ON: Her Majesty the Queen in Right of Canada. doi: https://horizons.gc.ca/en/2018/10/19/the-next-generation-of-emerging-global-challenges/

Seligman, M. E. P. (1975). *Helplessness: On depression, development, and death.* San Francisco, CA: W. H. Freeman.

3 Reconceptualizing 'Experience' for the Anthropocene

Kathleen Kesson

I came to higher education in the 1980s a seasoned progressive activist with radical ideas about education; having worked with the University Without Walls network in the early 1970's to establish a center in collaboration with Native American activists from a variety of tribal affiliations in Oklahoma. We were immersed in the practices of experiential education, opting for "learning by doing"—and the doing often included political organizing and protest, as well as organic food production, Native American Studies, guerilla theater, alternative architecture, and other unconventional (at the time) elements of the emerging counterculture.

I thought getting a graduate degree would necessitate a bit of "radical soul-selling" but was pleasantly surprised to land in the welcoming arms of advisors who were critical theorists and innovative thinkers, who gave me leave to design my own master's and doctoral programs, who nurtured my insatiable curiosity and supported my freedom to go wherever my intellectual nose led me. Having read *Schooling in Capitalist America* by Samuel Bowles and Herbert Gintis, as well as Joel Spring's *Education and the Rise of the Corporate State*, both of which I had picked up at a library used book sale because I loved the titles, I needed little convincing to understand education as a tool of the powerful for reproducing the (unjust) social order that I had been protesting for much of my adult life. I was delighted to forage in the vast emerging critical literature of the time, which seemed to explain so much of what was wrong with the world.

One of my earliest AERA memories, perhaps in 1989 or thereabouts, was listening along with a large, rapt crowd to Maxine Greene. After her talk, I turned to the person sitting next to me, and said something to the effect that "I'm fascinated by Maxine's brilliance. But I recently read a book that has caused me to question the fundamentals of what she is saying."

"What book was that?" asked my neighbor.
"*Elements of a Post-Liberal Theory of Education*. I have to confess, it has made me question everything, especially those sacred icons of

liberal educational thought I have been reared on. Dewey, Freire, even Carl Rogers. If I listen to Maxine now, I hear her devotion to modernism, growth, progress, the individual—concepts that I am increasingly beginning to question."

"Your name?" he asks.
I tell him, and he reaches for my hand to shake it.
"And yours?" I ask.
"Chet Bowers."

I recall his self-effacing grin, my embarrassment, and his generous allowance for my naiveté. We became friends on the spot. For a few years I read everything Chet wrote; I can no longer make that claim as he is probably one of the most prolific academic writers of the 20th and 21st centuries. When curriculum historians and other social theorists look back to these times, he may come to be seen as the most misunderstood, misrepresented, provocative, persistent, and prescient educational thinker of the current era. Chet Bowers suffered neither fools nor poseurs, and feared no critical engagement with his ideas, thus setting himself up for a lifetime of rabid, often ad hominem attacks, especially from the leftist edge of the educational theory spectrum. He let none of us off the hook: As he outlines in "The Cultural Dimensions of Ecological Literacy" (1996), most modern educational thinkers, whether technocratic, romantic, or emancipatory, base their prescriptions for education "on the same cultural assumptions that gave legitimacy to the Industrial Revolution and, now, the emerging Information Age" (p. 6). During a period of theorizing when few Western scholars questioned the centrality of the autonomous, rational self in their theorizing and how this conception of what it means to be human relates to the accelerating degradation of the environment, Chet was sounding the alarm over the inability of education scholars to examine their taken-for-granted assumptions about the nature of human existence, language, knowledge, and thought. In so doing, he laid the foundation for a philosophy of education relevant to what many scholars are calling the Anthropocene, the current geological era defined by the enormous and destructive human impact on Planet Earth.

The Crises of the Present

Readers of this book do not need reminding of the litany of crises we face: The extinction of over a million plant and animal species, the pollution of the oceans with plastics, the melting of glaciers, the rising of the seas, and climatic changes bringing increased numbers and intensities of fire, flood, and winds. The Earth is in revolt. Those of us who possess the will and the capacity to face reality cannot but be concerned about the dislocations, famines, and loss of life and property already happening, with more to come in the future. The temporal details of the Anthropocene are contested—some theorists cite its beginning almost 12,000 years ago at the inception of human agri/culture, some mark its start with contact between the "new" and "old" worlds after 1492, some with the invention of the steam engine by James Watt in the 18th century, and others with the first nuclear weapons test in 1945 (Royle, 2016, para. 14). Whatever marker one chooses, and the stratigraphic jury is still out on this, what is most important is that our generation—those of us born in the mid-20th century and forward—may be responsible for the elimination of life as we know it on our planet. We knew what was happening and did little to stop it. Nor did education scholars, responsible for preparing the next generation for life on a fragile planet, begin to seriously question the foundational assumptions of our educational theories, policies and practices, other than some scattered "posthumanist" discourses that have to date had little actual effect on either the content or the processes of schooling.

While action on all fronts is urgent, education is a primary vehicle for cultivating people with the knowledge, dispositions, and skills to survive and thrive in the coming era. Education could help to bring about the change of consciousness necessary to cultivate this "new human," but what this means is a radical rethinking of current theory, practice, and policy. James Moffett (1994) noted that:

> The generation about to enter schools may be the last who can still reverse the negative megatrends converging today. In order for these children to learn the needed *new ways of thinking* the present generation in charge of society must begin to set up for them **a kind of education it never had and arrange to educate itself further at the same time**.
>
> (Emphasis added, p. xii)

If Moffett's (1994) vision was correct, of course, it is by now too late to reverse these "negative megatrends" even if it were feasible for us to "set up for them a kind of education (we) never had and arrange to educate (ourselves) further at the same time." A bit like repairing the engine of an airplane while in flight. So, what to do? The question posed succinctly by American philosopher Roy Scranton (2018) is "we're doomed now what?" Can we make the shift from the Doomsday scenario in which we currently exist toward a future in which we learn to live within ecological limits, and perhaps in the process discover ways of living more satisfying and sustainable than the contemporary wage labor treadmill, materialism, and frenzied consumerism we find ourselves in?

While the many problems we face are human caused problems and are thus amenable to human solutions, we would be naïve to not recognize the powerful human interests, forces and institutions that are committed to maintaining the status quo (even if, to rational thinkers, this seems suicidal). I agree with those who say that there is a fundamental error in attributing the current planetary ravages to humans in general, disregarding the reality that it is a specific civilization that has driven these processes of resource extraction, labor exploitation, capital accumulation and what we can only call "ecocide." Colonialist civilization and its centuries long narrative of conquest, genocide, plunder, slave labor, and economic imperialism have created the conditions of this new age that we call the Anthropocene, and which some scholars suggest we more rightly call the "Capitalocene" (see Moore, 2015). Chet called our attention to the commodification of culture, and in this, he is aligned with the critiques of capitalism generated by leftist scholars. But while most Marxists are primarily concerned with a more equal distribution of the fruits of human labor, Chet points us toward prioritizing the non-commodified aspects of culture embodied by indigenous people not in the thrall of global market ideologies.

In his identification of high-status knowledge—the "experimental orientation of modern cultures" (Bowers, 1996, p. 9)—and low status forms of knowledge, he reminds us that "many traditional cultures developed highly complex patterns for insuring that the wisdom accumulated and refined over generations of dialogue between humans and their environment was passed on to future generations" (Bowers, 1996, p. 7). It is his recognition of local and place-specific ways of knowing, of the interwoven fabric of intergenerational knowledge, and the importance of honoring traditional ways of knowing which preserve the ecological commons that draws the ire of critical theorists, who associate these ways of knowing, as do most modern people in industrialized cultures, with "primitive" forms of culture.

The Meta-Narratives that Shape
Our Knowing and Being

A question central to Chet's thinking about cultural and ecological literacy is this: "Why has ecological literacy become so difficult for modern cultures to understand and carry out successfully?" (Bowers, 1996, p. 7). A meta-narrative, sometimes referred to as a "worldview," is an overarching story that purports to explain how the world works, and provides a big picture or pattern into which micro-events and stories can fit. The grand narratives that shape contemporary life in the West emerged with European Humanist philosophies, which liberated humanity from much of the superstition and irrationality of the medieval Christian Church and initiated an era of scientific thought and rationality. The story of humanity in much of the "developed" world since then is a narrative of progress, of increasing individual freedom and rights, of economic growth, constantly improved standards of living, and the capacity of rational thinking to solve all human problems. The advent of empirical science and its concomitant technologies created a constellation of knowledge and power that has subsumed alternative narratives, though even in WEIRD (White, Educated, Industrial, Rich, Democratic) societies, religious narratives co-exist alongside scientific and secular ones. The capability of the modernist mental model to capture our imaginations and shape what is possible seems intractable; the tropes and metaphors that constitute our thinking and being are embodied in the institutions that govern our lived experience: Large capitalist corporations, dominant religions, the (in)justice and incarceration systems, the militaries, the media and advertising machines, and perhaps especially, the schools, so much so that most people take them for granted. As Bowers so clearly reminded us, the root metaphors in our languages encode "earlier shared understanding and patterns" and thus "reproduce the moral templates that are to be used in different relationships" (Bowers, 1996, p. 8). These taken-for-granted concepts are woven into our everyday communication and lives, and few people question the tacit assumptions inherent in the language they speak.

Education can change our mental models and thus shift our narrative—but only if educators become conscious of the ways they pass on outmoded ways of thinking that no longer serve the survival interests of life on a small and fragile planet. And education cannot exist in a vacuum; new forms of learning must develop in tandem with new ways of thinking about economy, ecology, forms of social organization, and relations between people and more-than-human (see Abram, 1997) forms of life.

I believe Chet was pushing us to acknowledge the ways that deepening our understanding of indigenous worldviews, so radically different from our modern, Western narrative, might point us toward the kind of thinking we need to cultivate if we are to survive and thrive in this new era. We need, says Naomi Klein, to "articulate...an alternative worldview to rival the one at the heart of the ecological crisis" (2014, p. 462). Such an alternative worldview would need to be "embedded in interdependence, rather than hyper-individualism, reciprocity rather than dominance, and cooperation rather than hierarchy" (ibid.). We can look to indigenous scholars and educators for an elaboration of these themes of interdependence, reciprocity, and cooperation (see Cajete, 1994).

Worldviews are highly resistant to change, even when we recognize that they no longer serve as useful roadmaps for organizing our experience. It is only when some crisis (or in this moment, multiple crises) forces us to recognize how maladaptive our worldview has become, that are we prepared to consider something new. Our educational policies and practices are steeped in a modernist worldview that has outlived its usefulness, and which in fact is leading us to the brink of destruction. A philosophy of education equal to the task of guiding us into and through the Anthropocene must, therefore, reconceptualize its most fundamental assumptions about what it means to be human and how we come to know. While Bowers has pointed us in the right direction, we must all undertake the arduous journey of enacting more appropriate ways of thinking and being. A place to start is with the reconceptualization of what it means to be an "I."

Rethinking Ontology

I-ness
but a brief moment in a
stream of becomings
perceptions mingle with feelings
a chill wind carries
a sense of dread
the call of a bird lifts our heart
ocean waves
draw us to the horizon of enlightenment
premonitions, prehensions, visions
compose the inside/outside nature of experience.

Once known as metaphysics, ontology is the study of what it means to be human, including the broad categories of being, becoming, existence, and meaning. One main idea that has profoundly shaped the Humanistic assumptions that our modern world is based on, and especially the science of psychology, upon which much of educational theory is constructed, is that of the separate self. The notion of the individual, the "I" as an entity bounded by skin and personal perceptions, motives and behaviors, surrounded by stable substances and objects in space that constitute separate "others" to manipulate, utilize, and transact with, is a flawed construct, one that developed in the context of the Western Enlightenment along with the subjugation of nature and the application of reason and logic to all of the problems of existence. Western mythopoetic narratives, says Bowers (1996), "represented humans as separate from nature—as being in control of their own destiny regardless of how their actions degraded the environment" (p. 7). This sense of separation, mastery, and control in concert with an economic system predicated on resource extraction, endless growth, and needless consumption has led us to the ecological tipping point at which we find ourselves. We must therefore cultivate an ontology that is *relational*, that understands there is no distinct separation of self and other, of knower and known, of subject and object, but rather endless flows of being and becoming with which we are deeply interconnected with everything in creation, visible and invisible, substantive and molecular, objective and subjective (Kesson & Oliver, 2002). Few Western philosophers, other than Alfred North Whitehead and Gregory Bateson, understood the ways in which such complex moments of experience (Whitehead called these "occasions" [Oliver, 1989]; Bateson called them the "patterns that connect" [Bowers, 1997, p. 149]) come to enjoy some level of persistence over time and space, accounting for our sense of the apparent stability of self and the world. Buddhist concepts concerning the transient nature of phenomena, the arising and passing away of forms, come closer to a relational ontology, as does the Tantric notion of "emanative flows" interacting and interpenetrating in an "eternal dance of the macrocosm," the "unending cycle of evolution and dissolution" the Yogis call Brahma Cakra (Shambushivananda, 2017, p. 187). In the context of relational being and becoming, virtually all aspects of education require reconceptualization: Everything from our notions of individual achievement and personal empowerment to our valuing of independence and autonomy, from our theories of human development and cognition to theories of experience. If everything is in process, and relational, then we must awaken to the profound interdependence between the human organism and the environment, the life histories and trajectories of "objects" and our own implication in these. A philosophy of education for the Anthropocene would embrace this multidimensionality, the whole of ontological experience.

This will be no easy task; as Bowers noted in so much of his work, "the modern way of understanding ourselves as individuals is so deeply held, and influences so many aspects of modern cultural life...that it is nearly impossible to think in any other way" (1997, p. 145). To fail in this, however, to "continue to reinforce the modern image of the individual as the basic social, psychological, and moral entity" (ibid.) will only result in "furthering the double bind of basing educational reform on the cultural ideals that equate progress with exploiting the environment" (1997, p. 146). In suggesting to us the need to reconceptualize how the modern form of individualism is understood, Bowers foregrounded the importance of "taking seriously the ecologically sustainable forms of knowledge, values, technological practices, and sense of community that characterize cultures that have resisted the reductionism and consumer orientation of Western modernity" (p. 146), thus turning us from ontological to epistemological considerations.

Epistemological Pluralism

Epistemology asks fundamental questions about the nature of knowing. *How is knowledge constructed? What are the sources of knowledge? How do we come to know anything? How can we know what is true?* Western science, from the 15th century on, with its claims of objectivity and universality, has become the epistemological *sine qua non* of the modern world. In the process of valuing a particular version of scientific investigation and reason over all other forms of knowledge construction, and in the context of conquest, patriarchy, colonialism, and economic imperialism, ways of knowing that exist outside these contours have been marginalized or suppressed: embodied knowing, intuitional knowing, narrative knowing, aesthetic knowing, mythic knowing, and intergenerational knowing, to name a few.

In the Western scientific/Humanist paradigm, there is the idea that the human intellect is capable of generating timeless truths, independent of context. One consequence of this idea has been the misapplication of knowledge formed in one context to other contexts. For example, when Western agricultural experts exported their "Green Revolution" in the 1960's in an effort to bring an end to global starvation, they failed to take into account local and contextual factors, resulting in the loss of biodiversity and the impoverishment of small farmers all over the world. Just as in agriculture, Western monocultural models of education have spread across the planet, resulting in the loss of language, tradition, culture, and indigenous ecological knowledge. Some scholars have aptly called this "epistemicide" (Santos, 1998).

As Chet so rightly taught us, knowledge is not a "thing-in-itself" that can be transmitted from one isolated mind to another, or from a digitized environment to a human brain, via language or image (Bowers, 2000). Knowledge is part of an ever-changing system, a pattern of relations that is embedded in culture, and meanings are metaphorically encoded in language. The root metaphor of mechanism, for example (a theory established in 17th century Europe that understood the world as a machine) has "influenced western approaches to medicine, architecture, education, agriculture—not to mention our language and thought patterns" (Bowers, 2002, p. 4). We transmit worldviews and taken-for-granted cultural habits with every word we utter. We need to cultivate understandings of the way that language shapes how we perceive and understand the world. Robin Kimmerer, author of *Braiding Sweetgrass* (2013), speaks of the "grammar of animacy" in her Potawatomi language and contrasts that with the English language, in which only humans deserve specific pronouns, while all other animate and inanimate entities are classified as "its." In Potawatomi, "(l)iving beings are referred to as subjects, never as objects, and personhood is extended to all who breathe and some who don't" (2017, para 6). This is just one of many examples of how the very words we use construct how we understand the world. Bowers (1996) summarizes the importance of this insight for educators:

> When framed in terms of the need to conserve cultural patterns that are ecologically sustainable, this understanding of the culture-language-thought connection radically changes how we think about the content of every area of the curriculum. (p. 9)

How might educators begin to teach science as if the Earth was a sacred, living organism, rather than a machine? How then, might this affect the worldviews of our students? And how, in turn, might this affect the life choices they make? A relational epistemology asks more of us than that we simply "teach" or "acquire" neutral facts. To truly know anything, in a deep way, we must embrace the occasion of knowing in its temporal multiplicity: Understanding the past (how the knowledge was made) the present (what does it mean to me in this moment), and the future (what are the consequences of this knowing?). Conceptions of knowledge (where it comes from, how it is made, how it is transmitted, and what it means)— are intimately related to culture, cosmologies, histories, religious beliefs, and values. The "ecocide" that now structures our planetary experience is related to the "epistemicide" that resulted from the violent triumph of Western culture and capitalism and the marginalization of the vast diversity of ways of knowing that have evolved on our planet, many of which could provide justification and support for an epistemology that is ontologically relational, which recognizes and values the more-than-human world, and that understands creation as sacred, not merely utilitarian. To heal these deep epistemological wounds, we modern people (those of us raised in WEIRD societies) will need to expand the boundaries of our sources of knowledge: what might it mean to discard a notion of an "us" who think and a "them" that do not? Can we learn to "think like a tree?" (See, for example, Wohlleben, 2015.)

Cultivating a Theory of Experience for the Anthropocene

Dewey's Theory of Experience

John Dewey is one of the central thinkers of the Western Enlightenment, and an early target of Bowers's critique (1987). Dewey's important "theory of experience" (1938) sets forth a specific understanding of the human subject (ontology), and how they go about the task of knowing (epistemology). As we change the story of what it means to be and to know, we can then begin the arduous task of reconceptualizing how we go about the task of educating humans for this new era.

Dewey's theory is an educational ideal which has attained the stature of dogma with teachers and students alike, who when asked about their preferred method of education will often cite some version of "hands-on" learning. Experiential and student-centered learning are valued concepts in the educational world I inhabit in Vermont, where perhaps the most progressive (in a Deweyan sense) educational legislation in the country has been enacted. Here, educators are expected to help students integrate their interests into a "personalized learning plan," reflecting Dewey's concern with meeting the young person where they are in order to connect them to organized subject matter (Dewey, 1902/2011). As well, there is a strong emphasis on expanding the walls of the school to include the community with its greater variety of experiences available, reflecting Dewey's concerns for the meaningful association of human beings (Dewey, 1916/1966). And in this place of great natural beauty and robust commitment to sustaining the working landscape much of both the formal curriculum as well as student-driven learning choices focus on environmental, agricultural, and community issues, which could be construed to reflect Dewey's interest in the organic ties between the human organism and the environment, a relationship in which change is reciprocal (1929). In terms of responding to the ecological crisis, Vermont education is probably miles ahead of most places in the United States.

Dewey wrote *Experience and Education* (1938) to address the many misunderstandings and misapplications that had arisen concerning his ideas about education. In it, he set about to formulate criteria for discriminating between experiences that are educative and those that are miseducative. In the book, he develops the ideas of interests, initiative, purpose, motivation, transaction, continuity, and growth, and discusses the more natural forms of social control that might evolve when artificial forms of authoritarian control are removed from the educational environment. He emphasized the necessity to link experience to more organized subject matter, a task that he felt progressive educators had neglected.

Dewey's theory of learning from experience embodies a deliberative rational process in which problems are posed based on the emergent interests of the student. Once problems are identified, the student sets about to plan a solution, whether the problem is social, environmental, or artistic. To solve problems, one first observes the relevant conditions, and then forms hypotheses about possible solutions. In this process, the comparative values of differing choices are assessed and determinations are made about potential ways to reach desired ends. The order or sequence of activity is planned, the student takes action, and then evaluates the success or failure of the action, at which point the learning cycle may begin again, sparked by whatever new problems arose from the preceding cycle. This process, commonly known by the phrase "constructivist learning," is a tidy, conceptually sound, linear (if vaguely recursive) model of how one comes to know something, and by extension, solve a problem-situation.

Bowers' Critique of Dewey

There has been some effort by scholars to cite Dewey's "evolutionary naturalism" as an indicator of his importance to ecological thinking (see, for example, Hickman, 1996), based in part on his notion of intelligence as resulting from human transactions with the environment. The fact that he "made problematic situations the starting point and central focus of experimental inquiry" (Bowers, 2003, p. 1) and that the crisis-ridden moment offers us no shortage of problems to solve, coupled with his understanding of problem-solving as a communal event, might suggest his continued relevance as an ecological/educational thinker. Bowers thought otherwise, for the following reasons. Dewey's model of how people learn, and what kind of knowledge is most valued privileges a number of Enlightenment/Humanist assumptions that can now be seen as counter-productive to developing the ontological and epistemological dispositions necessary to create the conditions for survival in the present era. In Bowers' critique of Dewey's articulation of constructivist learning (2005), he highlights three conceptual failures on Dewey's part:

- The failure to appreciate epistemologies outside the modern, Western frame.
- The failure to recognize the value of intergenerational, traditional knowledge that enables sustainable cultural practices.
- The failure to value cultural traditions that resist modern, commodified lifestyles.

For Dewey, the rational individual and their inquiry-based, experimental relation to established (scientific) knowledge was the exemplar of modern thinking. This privileging of a certain way of knowing over other epistemologies, the failure to take cultural differences in ways of knowing and the associated "approaches to community and human/Nature relationships" (Bowers, 2003, p. 3) into account highlights the sterility of the theory. Compounding his epistemological prejudices is his tendency to label anything not-modern as "savage" or "barbaric," clearly demonstrating his predisposition to a linear model of evolutionary progress, a mode of thinking that failed to anticipate the disastrous consequences of Western, consumer-driven forms of growth and development. If Bowers is correct in his assessment, is there anything worth preserving in Dewey's theory of experience that might be of relevance to educators today?

Resolution? Reconciliation?

Both Bowers and Dewey shook up our notions of the atomized individual – Dewey with his end–in-view of social cooperation and community life (1956, p. 16); Bowers with the Batesonian notion of the "patterns-that-connect" (Rengifo et al., 2011). It's worth noting that Dewey did recognize the importance of knowing not just a thing, but its connections:

> A wagon is not perceived when all its parts are summed up; it is the characteristic connection of the parts which makes it a wagon. And

these connections are not those of mere physical juxtaposition; they involve connection with the animals that draw it, the things that are carried on it, and so on. (1916/1966, p 143)

And importantly, he reminded educators of the critical need for engagement with actual objects and events, noting that when mental activity is separated from active concern with the world (*experience*) words and symbols come to take the place of ideas, and:

> We get so thoroughly used to a kind of pseudo-idea, a half-perception, that we are not aware how half dead our mental action is, and how much keener and more extensive our observations and ideas would be if we formed them under conditions of a vital experience which required us to use judgment: to hunt for the connections of the thing dealt with. (1916/1966, p. 144)

Clearly, both Dewey and Bowers valued real-life concrete learning over cognitive abstractions as a basis for developing intelligence. As Dewey so wisely proposed, books and words and ideas should *supplement* actual experience, not *substitute* for it. And Bowers was firm in his belief that the printed word "promotes abstract thinking that leads to separating ideas and policies from different cultural contexts" (Bowers, n. d., para. 3). However, their rationales for rejecting the authority of subject matter differed. While Dewey rejected the transmission of knowledge created in the past in favor of young people developing their own ideas based on their immediate perceptions and reflections, Bowers did not reject the authority of inherited knowledge as much as he worried about the "pervasive influence of the inherited vocabularies on all human activity" (page unknown, ibid.) For Bowers, it is not knowledge of or from the past in general that is problematic—he values intergenerational/traditional knowledge—it is a *particular* past that concerns him; a post-Enlightenment, Humanistic past that privileges the autonomous human, notions of linear progress, human separation from nature, and the myth of objective truth, conceptions that have led us to the ecological tipping point.

Perhaps their instincts were not totally antithetical. Yes, Dewey believed that traditional beliefs and inherited values were signs of unintelligent thinking, and an invitation to oppression. He did not reject knowledge of the past altogether, however; in *Experience and Education* (1938), he poses the question: *How shall the young become acquainted with the past in such a way that the acquaintance is a potent agent in appreciation of the living present* (p. 23)? Here these two thinkers might agree that nothing could be more important in the living present than young people deeply understanding the connections in their world—the connections of what they eat to the health of the land, of what they buy to carbon emissions, of how they spend their time to the well-being of their communities…and that their education should involve engagement in substantial experiences that sharpen their intelligence, engage their emotions, and awaken their judgment.

Perhaps they would agree that it has never been more important to place concrete life experience at the center of how we educate this new generation. For a long time now, young people have been subjected to a passive form of education, receiving "high status" knowledge (abstract, decontextualized information) from textbooks and tiny screens in hopes they will become literate and numerate. Some young people do succeed at this, but a large number of them become disengaged, drop out, or turn to drugs and other forms of stimulation to remind them that they are alive. It is time to unhook from reliance on textual and digital environments and allow for their immersion in the sensory material world, so that they can come to know their interconnectedness with trees, soil, sea, ice, and weather. They need to hear the calls of birds (those that are left) and learn their names, and study the tracks that possum, raccoon, and deer make through their habitats. They need to be engaged in "re-inhabitation"—planting trees, feeling a part of the solution that will come with regenerating our forests, while at the same time, learning the indigenous histories of how and by whom the land has been inhabited, and what forms of past violence and exploitation exist in its "living present." They need to get their hands in the dirt and understand the interdependence of soil, weather, seeds, and rain, even if this dirt is on urban rooftop gardens, and appreciate the labor that has gone into feeding humanity through history. They need to be engaged in creative invention, learning how to apply mathematics and practical engineering in eco-centric ways that lead to a sustainable, livable future. And perhaps most important, they need first-hand, in-depth experiences with non-commodified culture such as singing, dancing, painting, crafts, inventing, sculpture, cooking, drama, games, and sports, activities that enable relief from the addictive consumer culture in which they have been raised. Face-to-face engagement with a variety of adults is essential. And when has it ever been more important to cultivate networks of mutual support and solidarity? Yes, there are young people who might lack the physical strength or ability to engage in robust sorts of activity, and this needs to be taken into consideration. We need to create a world that draws upon the strengths and attributes of all of the people in a community.

Knowing, Being, and the Spiritual Dimension of Experience

The Anthropocene is upon us. This is not an abstract possibility, but a concrete reality, and life on earth is changing rapidly for many, many people. *Catastrophe* has become a common *experience*. Many adults now say they would not have had children had they known what was coming. But children keep being born, and we need to prepare them to meet the uncertainties of the world that is upon us.

Bowers was not overtly engaged with notions of spirituality—at least in his early work. And I did not spend much time in his company in later years, so he may have changed or enlarged his perspective on this. However, if we attend carefully to his support for culturally diverse knowledge systems, especially those grounded in intergenerational communication that acknowledges the interconnectedness of all species and the importance of sustaining the viability of the natural systems on which life depends, we must embrace the possibility of forms of "wisdom older than (the) thinking mind" (Abram, 1995, p. 313) and the resonant presence of numinous and mysterious elements of nature from which Western thinkers, in their modern, autonomous ways of knowing, have split.

In a relational ontology, the "ground of information" from which we might draw is lush and fertile: The vast web of material interconnections in the human/nature ecology; the feelings and emotions that arise when we come into contact with a human or more-than-human other; the prehensive strata of consciousness, where the visible and invisible worlds engage in a quantum dance of coming-to-know (Kesson & Oliver, 2002). Epistemological pluralism would draw upon a more robust repertoire of ways of knowing to solve problems: Narratives and myths that might connect us to ancestral wisdom; intuitional insights not necessarily grounded in empirical data; practical considerations decoupled from notions of forward movement and progress. Only recently have we begun to understand the wisdom encoded in more traditional, ecological, intergenerational ways of knowing ("low status" knowledge) that are embedded in the specificities of places, ways of knowing that have been marginalized or altogether displaced as the knowledge/technology formations of global capital have subsumed the cultures and habitats that might point us toward more sustainable life patterns. Bowers has been careful to note that this is not a "romanticizing" of indigenous cultures, or a wholesale adoption of premodern cultural practices. All cultures, modern and premodern, industrial and pre-industrial, colonial and colonized have their own unique inequalities and abuses, and our task is to cultivate epistemological pluralism, recognizing what is of value and what needs to be discarded in multiple forms of cultural knowledge. It is well beyond the scope of this paper to explore the many and complex dimensions of spirituality, religion, and culture; it must suffice for now to note the existence of vast unexplained depths of experience – in both what we call "internal" and what we call "external" experience— that science has not yet plumbed. In some ways, Western science can be said to have "disenchanted the world" (see Berman, 1981), successfully banishing magic, myth, and mystery from human experience. In other ways, technologies such as the electron microscope and time lapse photography have opened our eyes to wonders of creation formerly unperceivable by (most people's) ordinary senses. Perhaps those of us raised in WEIRD societies need a new metaphysics, one that does not discard empirical science and the awakenings it has brought, but one which also recognizes our organic embeddedness in complex, natural systems of human and more-than-human life forms, both animate and what we have perhaps mistakenly termed "inanimate." A relational ontology, with its blurring of the boundaries between self and other, subject and object, will require new consideration of the numinous and mysterious elements of nature and how we might, through care/full observation of and awakened participation with more-than-human others, manage to re-enchant the world.

Important to this aspect of human development is a deep understanding of the ways that ceremony and ritual play an integral part in how moral templates for sustainability are encoded and transmitted from generation to generation. In the WEIRD world, rituals are generally spectator sport or spectacles of individual achievement such as graduations. "Church" is effectively separated from "State" and any hint of the sacred is generally seen as a violation of this modern split (unless powerful people promoting dominant religions seek to collapse the boundaries). Meaningful rites of passage or markers of important events that might meet the criteria of a relational ontology are largely absent from our collective lives. People who lived closer to the land and its ecological imperatives, however, developed complex and sophisticated "social and spiritual technologies" (LaChapelle, in Devall & Sessions, 1985, p. 248) that maintained their *relationships* to the world around them (and it is telling that many indigenous cultural groups refer to non-human entities as "relations," while Western people generally limit the term to humans connected to them by blood or marriage). Rituals and ceremonies need not be connected to specific religions; elsewhere I have explored the ways in which the arts, and their capacities to foster transcendence and connection, can infuse the curriculum in ways that can build bridges between separate lives and communities, affirm a sense of the sacred, and confirm the human interdependence with the surrounding natural world (Kesson, 2004).

It is vital to be wary of cultural appropriation and the wholesale adoption of cultural practices we were not born to. Dominant monotheistic (Judeo/Christian/Muslim) religions are not reliant on place, they are transportable, and people who desire to affiliate can be converted. This is one reason why they dominate the globe; they have spread alongside the conquest and colonialism that subsumed so many of the local cultures of the world. Modern people of European descent cannot "convert" to indigenous spirituality, however, as the practices are specific to tribe or clan or community and often deeply connected to specific places.

People in WEIRD societies have to find their own ways out of the thicket of materialism and consumerism in which we have been imprisoned.

This is not a simple undertaking. I have been on this quest for a lifetime, and feel like I have only scratched the surface of what it might mean to be "post-humanist"—that is, understanding my deep connectivity with all species, decentering the needs and desires of my human self in my worldly actions, learning to hear, then listen to and heed more-than-human "voices." But I take to heart what I understand as Bowers' assumption of the inadequacy of the Modernist/Humanist project to foster a deep understanding of environmental limits and reconnect humans in meaningful ways with the non-human beings that animate the planet including animals, but also such entities as sea, sky, plants, moss, and rock. If we cannot do this, it is possible that we shall not survive. I hope this reflection has illuminated some of the ways that Bowers, in his prophetic work, pointed us not only toward a "post-liberal theory of education" but toward a "post-humanist theory of experience and education," capable of preparing young people with the knowledge, dispositions, and skills not only to survive, but to thrive in this uncertain new era we now call the Anthropocene.

References

Abram, D. (1995). The ecology of magic. Reprinted in T. E. Roszak, M. E. Gomes, & A. D. Kanner (Eds.) *Ecopsychology: Restoring the earth, healing mind* (pp. 301–315). San Francisco, CA: Sierra Club Books.

Abram, D. (1997). *The spell of the sensuous: Perception and language in a more-than-human world.* New York, NY: Random House/Vintage Books.

Berman, M. (1981). *The reenchantment of the world.* Ithaca, NY: Cornell University Press.

Bowers, C. A. (n. d.). Using ethnographies as a classroom strategy for introducing students to language issues, the cultural commons, and the limitations of print technologies. http://cabowers.net/pdf/Using_Ethnographies_as_a_Classroom%20Strategy_for_Introducing_Students_to%20Language_Issues.pdf

Bowers, C. A. (2005). *False promises of constructivist theories of learning: A global and ecological critique.* New York, NY: Peter Lang.

Bowers, C. A. (2003). The case against John Dewey as an environmental and eco-justice philosopher. http://cabowers.net/pdf/DeweysRelevance2003.pdf

Bowers, C. A. (2002). How language limits our understanding of environmental education. http://cabowers.net/pdf/howlanguagelimits2001.pdf

Bowers, C. A. (2000). *Let them eat data: How computers affect education, cultural diversity and the prospects of ecological sustainability.* Athens, GA: University of Georgia Press.

Bowers, C. A. (1997). *The culture of denial: Why the environmental movement needs a strategy for reforming universities and public schools.* New York, NY: SUNY Press.

Bowers, C. A. (1996). The cultural dimensions of ecological literacy. *Journal of Environmental Education, 27*(2), 5–10. https://doi: 10.1080/00958964.1996.9941452.

Bowers, C. A. (1996). The cultural dimensions of ecological literacy. *The Journal of Environmental Education, 27*(2), 5–10.

Bowers, C. A. (1987). *Elements of a post-liberal theory of education.* New York, NY: Teachers College Press.

Cajete, G. (1994). *Look to the mountain: An ecology of indigenous education.* Skyland, NC: Kivake Press.

Dewey, J. (1938). *Experience and education.* New York, NY: Collier Books.

Dewey, J. (1929). *Experience and nature.* New York, NY: Dover.

Dewey, J. (1916/1966). *Democracy and education.* New York, NY: The Free Press.

Dewey, J. (1902/2011). *The child and the curriculum.* Eastford, CT: Martino Fine Books.

Dewey, J. (1899/1956). *The school and society.* Chicago, ILL: University of Chicago Press.

Hickman, L. (1996). Nature as culture: John Dewey's pragmatic naturalism. In A. Light & E. Eric Katz (Eds.), *Environmental pragmatism* (pp. 50–72). New York, NY: Routledge.

Kesson, K. (2004). For the love of frogs: Promoting ecological sensitivity through the arts. *Encounter: Education for Meaning and Social Justice, 17*(1). https://kathleenkesson.com/for-the-love-of-frogs/

Kesson, K., & Oliver, D. (2002). On the need for a new theory of experience. In W. Doll & N. Gough (Eds.), *Curriculum visions* (pp. 185–197). New York, NY: Peter Lang.

Kimmerer, R. W. (2017). Speaking of nature. *ORION Magazine.* https://orionmagazine.org/article/speaking-of-nature/, New York, NY.

Kimmerer, R. W. (2013). *Braiding sweetgrass: Indigenous wisdom, scientific knowledge, and the teachings of plants.* Minneapolis, MN: Milkweed Editions.

Klein, N. (2014). *This changes everything: Capitalism vs. the climate.* New York, NY: Simon & Schuster.

LaChapelle, D. (1985). Ritual is essential (appendix F). In B. Devall & G. Sessions (Eds.), *Deep ecology: Living as if nature mattered* (pp. 247–250). Salt Lake City, UT: Gibbs Smith.

Moffett, J. (1994). *The universal schoolhouse: Spiritual awakening through education.* San Francisco, CA: Jossey-Bass.

Moore, J. W. (2015). *Capitalism in the web of life: Ecology and the accumulation of capital.* New York and London: Verso Books.

Oliver, D. W. (1989). *Education, modernity, and fractured meaning: Toward a process theory of teaching and learning.* New York, NY: SUNY Press.

Rengifo, G., Bowers, C. A., & Jucker, R. (2011). *Perspectives on the ideas of Gregory Bateson, ecological intelligence, and educational reforms.* Eugene, OR: Eco-Justice Press.

Royle, C. E. (2016). Marxism and the Anthropocene. *International Socialism, 151.* http://isj.org.uk/marxism-and-the-anthropocene/

Santos, B. S. (1998). The fall of the Angelus Novus: Beyond the modern game of roots and options. *Current Sociology, 46*(2), 81–118.

Scranton, R. (2018). *We're doomed. Now what?: Essays on war and climate change.* New York, NY: Soho Press.

Shambushivananda, D. (2017). *Thoughts for a new era: A neohumanist perspective.* Gullringen, Sweden: Gurukula Press.

Wohlleben, P. (2015). *The hidden life of trees: What they feel, how they communicate.* Vancouver, Canada: Greystone Books.

4 Beyond the Binary of Bowers

Jeff Edmundson

I first encountered Chet Bowers in the early 1980s when he participated in a mini-debate with Paulo Freire at Portland State University (PSU). I was a confirmed Freiran at the time, and, not able to hear Bowers' still-developing critique of Freire's roots in Enlightenment ideas of individualism and progress, I dismissed him as a weird kind of anti-left environmentalist who was no match for Freire.

I didn't come across Bowers again until 1995, when I began a doctoral program at PSU. Bowers had recently moved to PSU and I was urged by a fellow fan of Freire to take a class with him because he knew I was concerned about ecological issues. Bowers quickly designated me the class Red; because I used the word "capitalism," he baited me to offer the Marxist position on the topic at hand.

But I was hooked, to begin with, by his point that the language of progress–that progress is both linear and inevitable—was shared by both capitalism and socialism. Marxism, after all, is rooted in the assumption that capitalist accumulation creates the technological and economic conditions for the ideal socialist society. In the face of an already clear ecological crisis, Bowers offered an understanding that the cultural roots of our unsustainable society go far deeper than capitalism and that leftist approaches, including critical pedagogy, had paid far too little attention to the crisis.

Despite my red tinge, my research focus was transformed and Bowers eventually served on my dissertation committee. We went on to collaborate at conferences and other venues, where I focused on applying his theoretical work to the concrete reality of teacher education.

Cultural Bioconservatism

By the mid-1990's, most of the major components of Bowers' thinking had taken shape. In the 1996 article excerpted here, and for some years after, Bowers referred to his approach as cultural/bioconservatism. Though an awkward term, it in some ways better suited his work than the early-2000s concept of "ecojustice," which I understood as an attempt not so much to reach out to social justice oriented scholars as to emphasize that he too was concerned with social justice. Bowers never gave in to social justice scholars; rather, he wanted to demonstrate that his work encompassed their concerns and went beyond them. Arguably, his theoretical approach did not change significantly with the change to the more palatable label of ecojustice; both ecojustice and cultural/bioconservatism emphasize the centrality of thinking culturally and the importance of conserving sustainable traditions.

But some people couldn't get over the word "conservative," since the term connotes a determination to retain existing oppressions. Bowers consistently tried, without much success it seems, to challenge the conventional understanding of conservative by showing that free-market, individualist "conservatives" are really classical liberals; that is, their analysis starts with the assumption that the individual is the basic social unit, and the freedom of the individual is the paramount good (Bowers, 2003). The older meaning of conservatism, on the other hand, highlighted the unintended consequences of change and urged caution in changing social practices that had met the test of time— while also tending to ignore that in a hierarchical society those time-tested practices generally benefited the ruling classes. Bowers also wanted to distinguish his position from those who used the language of emancipation, which he argued were still operating within a liberal framework of both individualism and the assumption that progress is inevitable and always moving forward to something better. His labeling of these left-wing thinkers as "emancipatory liberals," of course, did not always sit well.

But my position has generally been that moving to a cultural/ bioconservative–or ecojustice–position doesn't mean giving up the insights of Marxism and critical theory. Class analysis and historical materialism still offer important ways to understand the modern world. For example, understanding how capitalist motives structure decision-making doesn't exclude further analysis from an ecojustice position. But Marxism doesn't adequately explain domination based in patriarchy and racism. Marxist approaches can show how capitalism exploits racism, but they don't really get at the cultural roots from which race-based domination springs. And Marxism especially doesn't explain the many-thousand-year-old assumption of human supremacy that leads non-indigenous humans to treat all the rest of nature as if it was ours for the taking. More broadly, most Marxism isn't conducive to understanding the way such ancient ideas are taken for granted in our everyday language and thus passed on unconsciously. As Bowers puts it, educators coming from a Marxist tradition "rely on a metaphorical language that encodes deep, culturally specific assumptions that are part of the Western Enlightenment" (2001, p. 33). Indeed, one of Bowers' central contributions was his relentless focus on the ways language structures our thinking. His point in the 1996 article excerpt that language "thinks us as we think within the language" will be further discussed later.

A Challenge to Critical Pedagogy

The understanding of the limits of Marxism and critical theory led some of us to ask if they could be seen from new angles. For example, Bowers (1996) quotes David Orr's dictum that "all education is environmental education" (1991, p. 90), I have found that the language of reproduction theory offers a path to connect that insight to left theory. Marxist-influenced scholars have for decades examined the ways that schools and other institutions in capitalist society serve to reproduce the hierarchies of class, race and gender. In a parallel manner (within culture), institutions such as the education system also reproduce the culture of unsustainability, expressed through assumptions about human supremacy, the inevitability of progress, and individualism–cultural beliefs that long predate capitalism. This is the essential point for those committed to a better world–to understand that the problem is not just capitalism, but the underlying culture. Certainly, capitalism has refined the disaster-making potential of the long-held assumptions, but these ideas can't be reduced to capitalism.

Similarly, critical pedagogy is too often trapped by the unseen consequences of some of its central language. For example, an excellent progressive teacher I know has an email signature with this quote from Freiran Joao da Veiga Coutinho: "*There is no neutral education. Education is either for domestication or for freedom.*" Let me acknowledge that such Freiran insights are mostly correct: Modern education *does* domesticate people in the industrial societies into accepting an unjust, unsustainable world. But, aside from the limitations of the "either-or" dualism, the back half of the statement highlights two key problems with much critical pedagogy: (a) The unexamined celebration of "freedom" and "emancipation"; and (b) the failure to consider education in non-modern cultures.

The implication of "freedom from restraint" directly reproduces the modern ideology of autonomous individualism, which has been so perfectly coopted by consumer capitalism. While critical emphasis on social justice appears to put the good of the community above the individual, Bowers repeatedly pointed out that individualism was reproduced by the critical theorists' emphasis both on freeing people from tradition as well as on centering the basis for knowledge in the individual's critical reflection, which, as he notes "foregrounds the authority of the autonomous individual and the modern idea that decisions should be made on the basis of the individual's immediate experience" (Bowers, 1995, p. 244).

Further, as Wendell Berry has so often noted, lack of restraint and lack of self-discipline leads us to wreck the ecosystems we depend on. Instead, Berry says, we should regain propriety, "an old-fashioned concept that I think is renewed by our present ecological concerns; it has to do with the question of how one should act, given one's place both in the world and in the order of creatures" (Grubbs, 2007, p. 21). Freedom rooted in propriety and restraint requires that one take responsibility for the consequences of one's actions.

And rather than education in indigenous cultures being framed with the dualism of domestication or freedom, ecojustice argues that, at least in sustainable cultures, education is neither, but instead should be understood as a passing on long-developed knowledge of how to live "right," that is, in ways that sustain the community within the limits of its bioregion. For example, Bowers quotes a declaration of Haudenosaunee elders: "We bring to your thought and minds that right-minded human beings seek to promote above all else the life of all things" (1995, p. 144), and then goes on to note that these words "encode the fundamental knowledge about the interdependence and fragile nature of the web of food, information, and spirit that we now call an ecosystem" (p. 145). People living sustainably are not "free" to use nature in any way they choose, and good education teaches that responsibility.

Or take the central Freiran concept of "banking education" (Freire, 1970). While banking education is part of an oppressive system in modern cultures, the passing on of knowledge in indigenous cultures looks, on the surface, a lot like "banking." Yet indigenous peoples don't invite their children to re-imagine the world, to find "their own" way of knowing–and to encourage them to do so is to colonize them into modern culture. Freirans sometimes made that mistake; Bowers co-edited a volume of writing by largely indigenous authors describing how critical pedagogy had undermined local cultures (Bowers & Appfel-Marglin, 2005). Bejerano, for example, found that Freiran Popular Education among Bolivian campesinos created hierarchies (of the educated vs the non-educated) that challenged the traditions of the non-hierarchical community where everything was based on reciprocity (2005, p. 62). Esteva, Stuchul, and Prakash suggest that Freire's focus on liberation left him "unable to perceive the victimization created by education.... He was unable to bring his brilliant critique of 'banking education' to the modern enterprise called 'education'" (2005, p. 29). At the least, Freirans need to acknowledge the limitations of critical pedagogy.

Yet as a teacher educator in the United States, I found it valuable—even necessary—to use the concept of banking education, because it helped create the "aha" that *how* you teach is as important as *what* you teach, that it encodes assumptions about hierarchy as well about learning.

An obvious resolution of this contradiction is to say that Freiran approaches can apply in modern cultures but not indigenous ones.

While that works as a stopgap, it doesn't resolve the underlying problem that critical pedagogy pretends to a universalism that is in fact peculiarly modern. A better answer is rooted in Bowers' understanding that an ecojustice pedagogy needs to distinguish between, indeed to focus on, which structures and beliefs need to be transformed and which ones need to be conserved (Bowers, 2003). This is explored below in the section on "Moving toward a Pedagogy of Responsibility."

Bowers and Critical Theorists

For Bowers, unfortunately, it seemed that to admit of any value in the work of Marx, Freire or even Dewey, was to surrender to the culture of modernity. Bowers was also unwilling to acknowledge that constructivist approaches to teaching were adaptable to ecojustice. Teacher educators can't ignore the rich psychological research showing that students construct understandings, but ecojustice thinking shows us that students do that construction with the cultural tools they have available to them— the language and conceptual frameworks that the culture provides. The job of ecojustice educators is to provide new conceptual tools so that students can construct new ways of seeing the world. This avoids the trap of reproducing individualism without ignoring the reality of constructive learning. For example, as a teacher educator I would introduce the work of Walter Ong (1982), who showed that oral cultures literally perceive the world differently than literate cultures–as literacy provides different conceptual ways to interpret and interact with the world. For example, Ong noted that without the abstractions created by literacy, oral cultures categorize objects much differently than literate cultures.

Perhaps the nadir of Bowers' conflict with critical theorists was the "legendary" fight in Educational Studies with Peter McLaren and Donna Houston. Responding to a paper about critical pedagogy, socialism and ecology (McLaren and Houston, 2004), Bowers (2005) first argued that Houston and McLaren–framed as representative of all Freirans— tended to use universalistic language that ignored the diversity of cultural knowledges systems. Bowers then turned to labeling McLaren and Houston as Social Darwinists because Freire indicated that cultures based on critical reflection are more advanced. While the point about Freire's older writing was technically accurate, since Freire was rooted in the Enlightenment and thus saw modern cultures as more advanced in certain ways than indigenous cultures, the Social Darwinist label grossly misrepresented both Social Darwinism and modern critical pedagogy. While Social Darwinism (itself a misuse of Darwin's work) took the culturally-created differences in wealth and technological development among societies and ascribed it to natural processes, critical pedagogy certainly understands the ways that systems of exploitation create the vast inequities in wealth.

Houston and McLaren (2005) countered by calling Bowers' work a form of "reactionary anti-imperialism" and implying he had brain damage. After reviving the tired trope that Bowers was a promoter of "Noble Savage" thinking, they pointed out that many indigenous cultures were already "contaminated" by the global culture, arguing that this legitimated the use of critical pedagogy. Yet they didn't try to offer any nuance regarding when critical pedagogy was appropriate.

David Gruenewald noted in an editor's statement his disappointment that what he and his co-guest editor had hoped to be a dialogue had been reduced to a battle. He was "saddened to see them insulting each other once again in print" (2005, p. 206).

At the time, I shared some of Gruenewald's concern about the lack of dialogue, though I was still firmly in Chet's camp. Similarly, I appreciated Gruenewald's widely read 2003 article "The Best of Both Worlds: A Critical Pedagogy of Place," which seemed like a major step forward at bridging the gap. Gruenewald brought together the discourses of critical pedagogy and place-based education, summarizing:

> A critical pedagogy of place aims to (a) identify, recover, and create material spaces and places that teach us how to live well in our total environments (reinhabitation); and (b) identify and change ways of thinking that injure and exploit other people and places (decolonization).
>
> (Gruenewald, 2003, p. 9)

Gruenewald went on to explicitly suggest that a critical pedagogy of place, drawing from the two traditions, would focus on both what in our lived world needs to be transformed, but also what needs to be conserved.

But Chet, unfortunately, fired back at Gruenewald, too. A critical pedagogy of place is an "oxymoron," he said, because "the linguistic tradition of relying upon abstractions .... fail(s) to take account of the intergenerational traditions of habitation that still exist in communities" (2008, p. 333).

Emphasizing the cultural commons (i.e., the knowledge passed down through generations in a non-commodified way), Bowers suggests that an ecojustice pedagogy would encourage students to explore "the differences between their experiences in various cultural commons activities and experiences in the industrial/consumer culture" (2008, p. 332). But this is precisely what Gruenewald was calling for. By bringing a critical and cultural view to place-based education, Gruenewald had corrected the ahistorical and apolitical aspects of some place-based thinking— and he was certainly deeply aware of the existing "intergenerational traditions of habitation" (Bowers, 2008, p. 333). Rather than challenging Gruenewald's critical pedagogy of place, Bowers might have framed his response as "adding to" or "extending." His emphasis on the existing cultural commons would have–and indeed has (see the discussion of a pedagogy of responsibility in the following section)—deepened the discussion rather than creating a wall between critical pedagogy and ecojustice. After all, recovering the commons and recognizing where it still exists are two parts of the same project. Gruenewald was using the language of critical pedagogy to say something very close to what Bowers had been arguing for years.

Unfortunately, Bowers was not going to acknowledge any value in the work of critical scholars. This was my direct experience in repeated conversations with him—he dismissed any attempt to find positive aspects in critical pedagogy. Chet's relentless disparagement of critical pedagogy and of any analysis that smacked of Marxism likely discouraged fruitful dialogue between critical pedagogy and ecojustice. Consequently, Chet's approach created walls that have been slow to crumble.

I can't count the number of people who have said some version of "Chet's right, but….," where the "buts" ranged from an overriding commitment to social justice teaching to irritation with Chet's interpersonal style. I've often noted that formulation and challenged it as excuse-making over the years–but there is a piece of truth in the statement. Chet taught us so much, but the impact of his work may have been limited because he was so driven to have his message heard.

Moving toward a Pedagogy of Responsibility

So, some of us continued to explore ways to bring the two fields into constructive dialogue. Rather than attacking Freire's notion of a pedagogy of liberation, Rebecca Martusewicz and I paid homage to it yet argued for a change of focus in offering a "pedagogy of responsibility" (Edmundson & Martusewicz, 2013; Martusewicz, 2019; Martusewicz, Edmundson, & Lupinacci, 2015), which urges us to attend to our ethical obligations to both human and non-human communities. What do these need to thrive? How can we enable that healthy life? This conception acknowledges the need for liberation from the chains of capitalism and domination, but not for unrestrained freedom; rather for a different set of bonds—ones built on the requirements of healthy ecosystems and communities. This way of seeing considers "just obligations," emphasizing the importance of obligations—responsibilities—to other beings, but adds that these obligations must be justifiable as non-oppressive and in service of a sustainable planet. Thus, for example, we are not morally free to pollute the air and water or exterminate other species; we are obligated to protect them as the basis of all life. Similarly, we are obligated to protect the health of human communities by maintaining or recovering practices which connect people rather than atomize them into self-seeking individuals. As Martusewicz

(2019) notes, "we are learning to ask, what does a good human economy in a healthy, that is whole, community look like? What does it require of us?" (p. 19).

Where critical pedagogy leads students to investigate and challenge the injustices created by capitalism, sexism and racism, and where Bowers' version of ecojustice focuses on the problems created by human supremacy, a pedagogy of responsibility ties these together by adding in the work done by ecofeminists. Karen Warren (1998) and others showed the connection between all forms of domination by exploring the underlying logic of domination –that is, once you accept that there are "superior" beings and "inferior" beings, it's an easy step to assume that the superior should dominate the inferior. And once you've justified one domination, it's a short step to accept other dominations. This understanding re-centers pedagogy to focus on domination, and also puts "ecology"—or the entire non-human world—on the same plane as the oppressions faced by humans.

It may not be surprising that teaching with a pedagogy of responsibility can look quite similar to teaching with critical pedagogy, at least in affluent industrial countries. But the differences are distinctions that make a difference (Bateson, 2000). The following examples highlight some of those differences by focusing on where a pedagogy of responsibility acknowledges the insights of critical pedagogy and ecojustice but takes them a step further.

Role of Language

Both critical pedagogy and the pedagogy of responsibility urge students to consider the role of language in shaping our consciousness. Critical pedagogy focuses on the ways that language can be used to persuade people to accept injustice—or to help them challenge injustice. A pedagogy of responsibility should do that, of course, but it looks deeper at the essentially cultural nature of language, the ways that "language [thinks us] as [we] think within the language" (Bowers, 1996, p. 9). In so doing, it focuses on understanding how language shapes and limits what we think, thus reproducing assumptions of an unsustainable culture—or how we can make language explicit in order to challenge those assumptions. For example, looking at language this way challenges the individualism that assumes we should encourage students to "think for themselves." It always creates discomfort for students—at all levels—to consider that they are not as free and unique as they think they are. Expanding on this idea, I would ask students to consider how to reformulate the request to "think for yourself," taking into account the bias in the phrase. They would come up with terms such as "think in this new framework," or "think in a non-anthropocentric way."

Further, using the lens of domination, it becomes clearer that human supremacy is in the same category as other domination-linked injustices, and thus of equal concern. And, in the ways that patriarchal language was until recently widely accepted as just the way things are, so, too, we can help students see how human supremacy is woven into our everyday language, whether it is seeing nature as a "resource" or in labeling as "animalistic" the torture of humans by other humans, or even the teacher leading a hike who challenges students to "conquer the mountain." To use the above language of critical theory, using such terms reproduces the domination of human supremacy.

From Examining Injustice in Our Own Lives to the Cultural Commons

Similarly, where critical pedagogy leads students to look at their own lives as a tool for challenging the injustice they experience, a pedagogy of responsibility can do that, but also asks students to look at their communities to uncover, conserve and reinvigorate the cultural commons—the non-commodified knowledge and skills that still exist, such as gardening, cooking and preserving food, art and music making. The importance of such non-commodified knowledge lies in the ways it embodies a way of living in the world that sharply contrasts with the modern world's reduction of everything to monetary value. While critical approaches also make this critique, a pedagogy of responsibility goes further by focusing not just on different relationships among humans, but on a different human relationship with the non-human world. In my Ecojustice Education classes, I first asked students to examine what commons skills they exercised. Then I assigned students to learn a new commons skill from a friend or relative. Students not only enjoyed learning something new, they valued the deeper relationships that resulted as well as getting a glimpse of more sustainable ways of living.

This discussion highlights the above mentioned attention by ecojustice scholars to what traditions are important to conserve. Where critical pedagogy focuses almost exclusively on "transformation," urging students to "rename the world" (Freire, 1970), ecojustice acknowledges that while some traditions need to be changed, equal emphasis must be placed on conserving the knowledge and traditions that have proven to be ecologically and socially sustainable ways of living.

Next, in the same way that critical scholars emphasize the intersectionality of race, class, and gender injustice (and where ecojustice mostly avoids intersectionality), a pedagogy of responsibility includes the oppression of the non-human world within its definition of intersectionality. The routinized abuse of animals, as well as workers, in factory farms is not just capitalist cruelty but is integrally tied in with the assumption that humans have a right to dominate the natural world.

Globalization

Finally, both critical pedagogy and the pedagogy of responsibility look at globalization. Critical approaches tend to focus on the colonization and exploitation of poor countries by the rich industrialized countries, but they often tend to see the solution as helping the poor countries catch up technologically. Even if the aim is just to have the wealth more equally distributed, this approach usually envisions all societies becoming modern industrial cultures. A pedagogy of responsibility challenges exploitation, of course, but challenges the assumption that industrialism (either capitalist or socialist) is the path forward. And, anticipating the argument that such ideas are just another example of the privileged determining what is good for the exploited, a pedagogy of responsibility draws from the work of scholars from the global South, such as Esteva and Prakash (1998), who show that much of what they call the world's "social majorities" actually resist unbridled technology and so-called "development," preferring to maintain ways of life that sustain both communities and ecosystems.

And, in contrast to the usual unquestioned push by progressives for the expansion of human rights, ecojustice and the pedagogy of responsibility join Esteva and Prakash in arguing that Western conceptions of human rights can be colonizing of indigenous cultures because rights language privileges the individual over the community, thus undermining the bonds that hold sustainable communities together.

Liberation or Care-Taking?

Most of the Freirans I know, focusing as they do on teaching for social justice, are also deeply concerned about the various ecological crises facing us—and they teach about that too. And, few of them would question that human supremacy is a concern (if a lesser one than patriarchy or racism), or that language shapes consciousness, or that individualism is corrosive. Similarly, few ecojustice educators would question the importance of teaching to challenge racism, sexism, and class exploitation, as these are inevitably entangled within environmentally destructive discourses.

This doesn't deny that there are real differences, or suggest that it's only a matter of what one emphasizes in one's work. Consider McLaren's recent call for unity among radical educators, which reasonably emphasizes the interrelatedness of different movements, and notably names Bowers among those who should be valued (2015). Yet in the same paragraphs he calls for a pedagogy of "total liberation," doubling down on the language of "liberation" without any recognition that it: (1) Leaves key questions undiscussed; (2) falls into the trap of the language of emancipation discussed earlier and especially, (3) carries the baggage of implying that we can be free of all traditions, all structures that limit us.

The obvious unanswered question for McLaren is "liberation towards what end?" He lists some admirable but generic examples of a more equitable world as well as one that is more ecologically sustainable—essentially an eco-aware democratic socialism. But what cultural foundations is such a society to be built on? If the foundations remain those of modernity, then they will perpetuate the anthropocentrism and faith in progress that have led to our cascading ecological crises. Such a culture will rely on the fantasy of unlimited growth on our finite planet, which we know has disastrous consequences.

Next, the language of emancipation is all too compatible with capitalism, which loves to liberate people from their traditions in order to better sell them products. As Bowers has shown so often, freedom in this culture is deeply linked to the belief that individuals should be autonomous, without understanding that autonomous individuals are unconstrained individuals. And as Wendell Berry has shown so well, unconstrained individuals are rapacious consumers and as such are irresponsible members of the planetary community.

Most importantly, a pedagogy of responsibility asks: In a world of total liberation, how can there be just obligations? A truly sustainable and just society must place significant limits on human behavior, as every successful indigenous society has done. The question is whether those limits are created within a context of domination, as exercised by all modern societies, or whether they are created based on natural limits (i.e., those that are needed to maintain living ecosystems) and based on maintaining healthy human communities.

And what does total liberation mean for indigenous cultures who already know how to live sustainably and equitably? Are they to join the modern world and be liberated from the ways of living that have worked well for so many generations? In reality, a "totally liberated" world within the frame of modernity would be shaped by the largely-unconscious assumptions—anthropocentrism, individualism, the faith in progress–carried by all people who have been educated in modernity.

Instead of "total liberation" as the slogan of a unified movement, a pedagogy of responsibility might suggest that it would be better to call for a total end to domination. Or better still, instead of totalizing language with its inherent universalism, let's call for a *"(re)turn to local care-taking."* Care-taking is meant to imply non-dominative relationships as well as the necessity of taking responsibility for the human and non-human communities around us, while "(re)turn" indicates both the need for modern societies to turn toward sustainable practices as well as for non-modern cultures to be able to maintain or return to ways of living which are informed by specific understandings of the local bioregion. Martusewicz, following Berry, emphasizes such care requires love, "not as a romantic notion, but as it is enacted in the forms of devotion and responsibility required by... life on earth" (2019, p. 18). Thus, she says, teachers and students "need to embrace specific sensibilities related to care and restraint, humility and kindness, to imagine what it could mean to live by these and to help others to live by them" (p. 7).

Conclusion

Tremendous steps toward justice have been accomplished under the rubric of critical theory. Still, having worked through critical pedagogy and while still valuing its insights, I remain convinced that Bowers' work and the scholarship influenced by Bowers, such as the pedagogy of responsibility, addresses fundamental issues not attended to by the various forms of critical theory. Ecojustice theory understands that the ecological crisis is at root a cultural crisis, that language such as "liberation" reproduces modern ways of thinking, from individualism to the contempt for restraint that undermines sustainable living. And, as explored earlier, recognizing the non-human world as equal to the human world compels educators to rethink many of our comfortable assumptions about who and what we are fighting for.

Though ecojustice and critical pedagogy are far from opposites, they both fit Bohr's definition of profound truths. A pedagogy of responsibility offers one path toward reconciliation between the two; perhaps even a synthesis. At any rate, it's essential that we engage in mutually supportive dialogue rather than academic one-upmanship. With the overwhelming crises upon us, it's far more urgent to work together for change than to fight with friends. Perhaps this book is one step toward that end.

References

Bateson, G. (2000). *Steps to an ecology of mind.* Chicago, IL: University of Chicago Press.

Bejarano, B. (2005). Who are the oppressed? In C. A. Bowers and F. Appfel-Marglin (Eds.), *Rethinking Freire: Globalization and the environmental crisis* (pp. 49–68). Mahwah, NJ: Lawrence Erlbaum Associates.

Bowers, C. A. (1995). *Educating for an ecologically sustainable culture: Rethinking moral education, creativity, intelligence, and other modern orthodoxies.* Albany, NY: State University of New York Press.

Bowers, C. A. (1996). The cultural dimensions of ecological literacy. *Journal of Environmental Education, 27*(2), 5–10.

Bowers, C. A. (2001). *Educating for eco-justice and community.* Athens, GA: University of Georgia Press.

Bowers, C. A. (2003). *Mindful conservatism: Rethinking the ideological and educational basis of an ecologically sustainable future.* Lanham, MD: Rowman and Littlefield.

Bowers, C. A. (2005). How Peter McLaren and Donna Houston, and other "green" Marxists contribute to the globalization of the West's industrial culture. *Educational Studies, 37*(2), 185–195.

Bowers, C. A. (2008). Why a critical pedagogy of place is an oxymoron. *Environmental Education Research, 14*(3), 325–335.

Bowers, C. A., & Appfel-Marglin, F. (2005). *Rethinking Freire: Globalization and the environmental crisis.* Mahwah, NJ: Lawrence Erlbaum Associates.

Edmundson, J., & Martusewicz, R. (2013). Putting our lives in order: Wendell Berry, ecojustice and a pedagogy of responsibility. In A. Kulnieks, K. Young, & D. Longboat (Eds.), *Contemporary studies in environmental indigenous pedagogies: A curricula of stories and place* (pp. 171–184). Rotterdam, Netherlands: Sense Publishers.

Esteva, G., & Prakash, M. S. (1998) *Grassroots post-modernism: Remaking the soil of cultures.* New York: Zed Books.

Esteva, G., Stuchul, D., & Prakash, M. (2005). From a pedagogy for liberation to liberation from pedagogy. In C. A. Bowers and F. Appfel-Marglin (Eds.), *Rethinking Freire: Globalization and the environmental crisis* (pp. 13–30). Mahwah, NJ: Lawrence Erlbaum Associates.

Freire, P. (1970). *Pedagogy of the oppressed.* New York: Continuum.

Grubbs, M. (Ed.). (2007). *Conversations with Wendell Berry.* Jackson, MS: University of Mississippi Press.

Gruenewald, D. (2003). The best of both worlds: A critical pedagogy of place. *Educational Researcher, 32*(4), 3–12.

Gruenewald, D. (2005). More than one profound truth: Making sense of divergent criticalities. *Educational Studies, 37*(2), 206–215.

Houston, D., & McLaren, P. (2005). The 'nature' of political amnesia: A response to C.A. 'Chet' Bowers. *Educational Studies, 37*(2), 196–206.

McLaren, P. (2015). *Life in schools: An introduction to critical pedagogy in the foundations of education* (6th ed.). New York: Routledge.

McLaren, P., & Houston, D. (2004). Revolutionary ecologies: Critical pedagogy and ecosocialism. *Educational Studies, 36*(1), 27–44.

Martusewicz, R. (2019). *A pedagogy of responsibility: Wendell Berry for ecojustice education.* New York: Routledge.

Martusewicz, R., Edmundson, J., & Lupinacci, J. (2015). *Ecojustice education: Toward diverse, democratic and sustainable communities.* New York: Routledge.

Ong, W. (1982). *Orality and literacy: The technologizing of the word.* London: Methuen.

Orr, D. (1991). *Ecological literacy: Education and the transition to a postmodern world.* Albany, NY: State University of New York Press.

Warren, K. (1998). The power and the promise of ecological feminism. In M. E. Zimmerman, J. B. Callicott et al. (Eds.), *Environmental philosophy: From animal rights to radical ecology* (pp. 325–244). Upper Saddle River, NJ: Prentice Hall.

5   Steps to an Ecology of Mindful Teaching in Outdoor Life (*friluftsliv*) Education

Revisiting C. A. Bowers and Arne Naess for a Deep Ecocultural Approach

Per Ingvar Haukeland

This chapter revisits a dialogue that took place in the *Trumpeter* in 1993 between Chet A. Bowers (1935–2017) and the founder of the deep ecology movement, the Norwegian philosopher Arne Naess (1912–2009) (Bowers, 1993b; Naess, 1993). From 1989 to 1991, I was a master's student at the University of Oregon with Bowers as my advisor. While writing my thesis in Norway in 1990, I began a collaboration with Naess, sharing an office at the Centre for Environment and Development, University of Oslo. During this period, I introduced Bowers to the ideas of Naess and vice versa, with a hope of a fruitful dialogue between them. It did not happen. Instead they spoke mostly passed each other. As Bowers critiqued Naess with his usual critical cunning eye on how taken for granted assumptions pass across generations, he did not fully grasp or engage with Naess' complex notion of the "ecological Self." And while Naess tried to find some common ground regarding how to understand the ecological crisis and the importance of individual and social responses, he did not fully grasp or engage with the language-thought-culture connection that Bowers advocated. I must admit I was disappointed, since I saw immediate correspondence between their ideas, which is the project I take up in this chapter.

I will relate the works of both writers to my interest in outdoor life (*friluftsliv*) education at the University of South-Eastern Norway. After a brief introduction to *friluftsliv*, the chapter unfolds in three parts. First, a critical appraisal of the works of Naess and Bowers on how to understand and deal with the ecological crisis. Second, an outline of a "deep ecocultural approach" that combines Naess' deep ecology approach and Bowers' critical ecocultural approach. Third, I apply this deep ecocultural approach in developing an ecology of curriculum development and mindful teaching in outdoor life (*friluftsliv*) education.

What Is *Friluftsliv?*

The Norwegian word "friluftsliv" (literally "free-air-life") is hard to translate. Sometimes, as in this chapter, it is used as outdoor life (Dahle, 1995; Gurholt, 2014; Gurholt & Haukeland, 2019; Henderson & Vikander, 2007). The White Paper definition says "*friluftsliv* is staying and being physically active in the free-air during leisure time with the intent of nature experience and a change of environment" (Ministry of Climate and the Environment, 2016, p. 10). This definition focuses on leisure, nature experience and "change of environment," not as environmental change, but as "change of scenery."

One Norwegian proponent of *friluftsliv*, Nils Faarlund is critical to this definition, since it does not say anything about caring for nature. *Friluftsliv* and nature-care are two sides of the same coin in his view. For him, nature is the home of culture and *friluftsliv* is a way home (Leirhaug, Haukeland, & Faarlund, 2019). Gunnar Breivik (1978) says there are two *friluftsliv* traditions in Norway: (1) The rural village tradition, linked to traditional culture, and (2) the urban city tradition, linked to urban people's recreation in modern culture. Breivik concludes that *friluftsliv* can be linked to both modern surplus forms of leisure (climbing, camping, walking, biking) and traditional self-subsistence forms of livelihood (hunting, fishing, foraging) (Breivik, 1978, p. 14). He says that both traditions are important expressions of *friluftsliv*.

If we simply translate "friluftsliv" with "outdoor life," "outdoor recreation," or "life in the open air," (Reed & Rothenberg, 1993) doing so does not fully capture how Naess understands it. Naess (1995) sees *friluftsliv* as a Norwegian root to the international deep ecology movement. He asks, "Has Norway anything to tell the world?," and answers, "I don't know anything else other than the classical Norwegian *friluftsliv*" (Naess, 1995, p. 15). According to Naess, "Outdoor recreation is often used for the activities more and more people in the industrial societies are engaging in during their leisure time." However, *friluftsliv*, he continues, is a clearer, more value-laden word that refers to the type of outdoor recreation that seeks to come to nature on its own terms: *to touch the Earth lightly*" (Naess, 1989, pp. 177–178). For Naess, it has something to do with how a person relates to nature. Nature is an abstract term, which means everything from a green area in the city to wilderness. Naess often uses the phrase "free nature" to express the importance of relating to a nature that is not controlled or domesticated by humans; it is beyond human control. Humans are also part of "free nature," but it extends beyond humans to what cultural ecologist David Abram (1996) calls "more-than-human."

In my view, *friluftsliv* can be studied as a three-fold phenomenon. First, as an *existential phenomenon*, it gets at how everyone seeks a balance between what imprisons and what liberates in between the self and its relations. The term was first used by the playwright Henrik Ibsen who, in a poem from 1859 ("On the Heights"), illustrated such an existential situation with a young hunter who found "*friluftsliv* for his thoughts" in the highlands (nature) and entrapment in the lowlands (village). I see this not simply as an engendered juxtaposition between nature and culture, but rather as an existential contrast between what at times makes us feel free and what makes us feel trapped. Second, as a *social and cultural phenomenon*, it changes with societal changes, from the traditional to the modern. According to the Statistical Bureau in Norway (SSB), as many as 9 out of 10 Norwegians say they engage in *friluftsliv* during the year, but the activities range widely. Third, as a *pedagogical phenomenon*, it encompasses environmental education, physical education, nature-connection and treading lightly in nature.

In kindergarten, the word "friluftsliv" shows up in the curriculum for "nature and environmental studies"; children shall experience nearby nature and learn about plants, animals, ecosystems, and the environment (Directorate of Education, 2017). While in elementary and secondary education, it is linked to the physical education curriculum (Directorate of Education, 2018); pupils are to be physically active in nature, to learn how to spend a night outdoors, navigate by map and compass, tread lightly and safely and in respect to nature, and partake in local cultural *friluftsliv* practices, e.g., hunting, fishing, foraging. In higher education, we find *friluftsliv* educational programs on the bachelor and master levels, predominately under the institutes of sports and physical education (Gurholt & Haukeland, 2019). The curricular aim in *friluftsliv* education, subsequently, has shifted from nature to physical education, and placed in a curriculum that includes competitive sports. Since *friluftsliv* is not its own subject, besides as an elective in upper elementary and secondary schools, teacher-students need to take physical education in order to teach *friluftsliv*. This makes a broader approach to *friluftsliv* in education more difficult. It may prevent students interested in the natural sciences, the social sciences and the humanities, as well as arts and crafts, to see *friluftsliv* as a teaching possibility.

The Ecological Crisis and the Role of Education

The prefix "eco" comes from the Greek "oikos," which can be translated as "home." The ecological crisis, hence, can be regarded as a crisis in the various aspects of what we can call "our home in life," including the interplay between nature, society, culture, community, and self. The ecological crisis is not simply a crisis of nature, but a social, cultural and existential crisis that can alienate us from our home. What role does education play in such alienation? How does it reproduce the problems with the ecological crisis and what does it do to deal with them?

Recent climate strikes among students worldwide, initiated by Greta Thunberg from Sweden, show two points that are wrong with our educational system: (1) It does not provide students with the education they need to cope with the ecological/climate crisis, and (2) it perpetuates the crisis by reproducing a way of thinking and acting that has led to a worsening of the problems of alienation, suppression, denial and distancing. It does not only do this, of course, and there is much positive engagement in schools and among the young, but the educational system with its curriculum rooted in Western culture does not help us fully understand what the problems are and how to deal with them, nor does it fully engage pupils to emotionally connect to nature and create strong relational bonds in which care and commitment can grow.

According to Bowers (1993a), education reproduces through teaching and curriculum the cultural tenets in the consumer-technological mainstream modern culture that perpetuates the problems with the ecological crisis. Examples of such tenets, according to Bowers (1996), are as follows: Change as progressive, i.e., always moving toward the better; anthropocentrism that places humans on top of the hierarchy of life; autonomy as the fullest realization of human potential and Western society as the most advanced and liberated of all cultures; science and technology as twin engines of human progress; and value as defined by market value. These taken for granted tenets, says Bowers, are reproduced unknowingly in curriculum and teaching.

Naess describes the ecological crisis as "an exponentially increasing, and partially or totally irreversible environmental deterioration or devastation perpetuated through firmly established ways of production and consumption and a lack of adequate policies regarding human population increase" (1989, p. 23). He also places the problem, similar to Bowers, in the context of the "techno-industrial consumerism" of a globalized culture that encroaches upon the health of Earth's ecosystems—our homes.

The problems with the ecological crisis also show up in *friluftsliv* education. The relationship between *friluftsliv*, the environment and sustainability has been studied in the Nordic countries from various disciplines (Fredman, Stenseke, Sandell, & Mossing, 2013; Gurholt & Haukeland, 2019; Hille, Aall, & Klepp, 2007; Sandell & Sörlin, 2000). Even the latest White paper on *friluftsliv* states that it "can give the individual a relationship to nature, which in turn can lead to an increased willingness to make environmentally sound choices" (Ministry of Climate and the Environment, 2016, p. 10). Whether this is the case has been scrutinized by climate researchers. One study found *friluftsliv* to be the third worst leisure activity for the climate that one can engage in (Hille et al., 2007). This is largely due to all the travel and consumerism of outdoor equipment and clothing. To travel far to see the last glacier, for example, contributes to the global warming that makes them disappear. This contributes to what is known in Norway as the "sustainability paradox" in *friluftsliv* (Gurholt & Haukeland, 2019). We may care for nature but how and where we do it may, directly or indirectly, harm nature through the impact on the climate. It is important, however, to point out that the carbon footprint is vastly different between *friluftsliv* activities, from roaming nearby nature to flying to the Alps to ski.

A cultural driver of the sustainability paradox can be found in the classic analysis of leisure by the Norwegian-American sociologist, Thorstein Veblen (1965). He described as early as in 1899 the emergence of a "leisure class" that was driven by conspicuous consumption and a pecuniary modern culture. Leisure is, for Veblen, a product of affluence caused by the modern-industrial-technological consumerist culture. Modern *friluftsliv*, hence, with its focus on "surplus leisure" may be paradoxically, a product of an unsustainable cultural development that has led to the ecological (climate) crisis. In traditional communities, nature was a necessity for survival, so to care for it was to care for oneself while in the modern society nature is more or less seen as something "out there," either as a bundle of resources or as an arena for recreation, that we feel we are no longer dependent on. The traditional and modern cultures represent two different forms of rationality and instrumentality; one is from within and the other is from without. Naess and Bowers seek, in their own ways, to find ways for people and communities to see nature as integral to culture, i.e., to get *back in*, so to speak. This requires, for Naess, an ecocentric approach to the web of life, which includes humans, of course, but not as superior to all else, but rather as a co-inhabitant.

As teachers of *friluftsliv* education, we need to be mindful of how the problems of the ecological crisis relate to education. If *friluftsliv* can instigate cultural change is perhaps farfetched, but it may be a contribution. It depends on what is done, how and in what context. Naess warned us back in 1989, "Extremely powerful forces are attempting to replace *friluftsliv* with mechanized, competitive, and environmentally destructive intrusions into nature" (1989, p. 180). He claims that the satisfaction of the need for outdoor life and the need for machine-oriented technical unfolding cannot take place *simultaneously*. At present, he says, "the socio-economic forces in the industrial countries are lobbying in favor of priority for the capital-intensive apparatus: the apparatus-poor life is a hindrance to 'progress'" (Naess, 1989, p. 178). A *friluftsliv* with simple means may be an apparatus-poor life, but more than this, it may be a life with rich ends (Haukeland, 2018).

Naess calls for a paradigm shift toward a nature-friendly *friluftsliv*. This may entail a revitalization of traditional knowledge and skills, to take a closer look at what is lost and what is gained in the modernization of society. An example that not all changes are progressive, as Bowers points out, is the race for the South Pole between Amundsen and Scott in 1911–1912. Amundsen learned from traditional cultures, such as the Saami and Inuits, and used traditional ski-makers for specialized skiis, dog sleds and used traditional clothing, while Scott used modern clothing, machines and ponies. The machines were quickly broken down by the cold and the ponies did not cope with the harsh environment. The choices were fatal for Scott and his crew. It is too easy to denounce all that is modern, but also too easy to renounce what is traditional.

Naess' Deep Ecology Approach

Arne Naess retired in 1969 as a professor of philosophy at the University of Oslo, only 57 years old, in order to devote his time to developing a philosophical response to the ecological crisis, which later became the deep ecology approach (Haukeland & Naess, 2008). This approach is rooted in what Naess calls "ecosophy," which he understands as "a philosophical world view or system inspired by the conditions of life in the ecosphere" (1989, p. 38). He encourages us to develop our own ecosophies in a way that bridges what we see as mattering in life and how we live our lives, as "all ecosophical insights should be directly *relevant for action*" (Naess, 1989, p. 37). However, he continues, "saying 'your own' does not imply that the ecosophy is in any way an original creation by yourself. It is enough that it is a kind of total view which you feel at home with...[and as] with one's own life, it is always changing" (Naess, 1989, p. 37). Even though an ecosophy is unique to the individual, it is also grounded in social, cultural and ecological relations.

Underlying Naess' deep ecology approach are three basic premises (Naess, 1989, p. 79). First, a human being is not a thing in an environment but a juncture in a relational system without determined boundaries in time and space. Second, this relational system connects humans, as organic systems, with animals, plants, and ecosystems, conventionally said to be either inside or outside the human organism. Third, our statements concerning things and qualities, fractions and wholes cannot be made more precise without a transition to field and relational thinking. Naess calls his own ecosophy for "Ecosophy T" ("T" for *Tvergastein*, the name of his mountain hut). It is outlined as a normative system with "Self-realization!" (with capital S) as the most ultimate norm (Naess & Haukeland, 2002). Naess speaks of both a "small self," as the "ego," and a large Self (with capital S) as the "ecological Self." The ecological Self co-evolves with the relationship to others we identify with both human and non-human. Identification creates, he says, intrinsic relations that are integral to who we are. The smaller self does not disappear as we extend into the larger Self. "We need not ignore or suppress the ego in order to broaden and deepen the self in contact with the Self" (Naess, 1989, p. 86). The ego and the Self reinforce each other. For example, Naess spoke often of a young Saami man that during the

Alta-dam demonstration in Northern Norway had chained himself to a bulldozer. When asked why, he said that "The river is part of myself!" A similar point was made by women in the Chipko-movement in India, who embraced the trees to prevent clear-cutting. To care for the river and the trees as they would for themselves makes ecosophically perfect sense.

The *deep ecology approach* is, according to Naess, an approach in the ecology movement that goes deeper into the causes and solutions to the ecological crisis. Also known as the "deep ecology movement," it is a social movement of people adhering to different ecosophies at the same time as they share the process of probing deeper into the problems and sharing their commitment for doing something about them. This creates the movement, and there are certain common ideas that its proponents share, which Naess and George Sessions formulated in 1984 in an eight-point platform (Naess, 1989, p. 29). The supporters of the deep ecology approach "tend to explain their eagerness to protect [nature] referring to their most basic views and attitudes, their value priorities, their understanding of what makes life meaningful" (Naess, 1995, p. 17). They base their approach on an ecocentric worldview of intrinsic relations that widens our care from the human to the more-than-human world.

When Naess speaks of *friluftsliv* as radical to the deep ecology movement, he understands it as "a way of life in free nature that is highly efficient in stimulating the sense of oneness, wholeness and in deepening identification" (Naess, 1995, p. 17). *Friluftsliv* can help us reconnect to our fellow creatures as co-inhabitants and "neighbors."

Bowers' Ecocultural Approach

The same year as Naess published his classic, *Ecology, Community and Lifestyle*, in 1974, Bowers published *Cultural Literacy for Freedom*, in which he applies a sociology of knowledge approach (after Schutz, Berger and Luckman) that "foster a dialectical relationship between the individual's own groundedness in cultural tradition and the capacity to envisage new social arrangements" (Bowers, 1984, p. 99). What evolved in the writings of Bowers is an ecocultural approach that addresses critically at the relationship, or lack thereof, between ecology and culture. His approach is rooted in his analysis of the metaphorical structure of language and how this makes up the conceptual and interpretive framework that guide, on a taken for granted level, curriculum development and how teachers teach. Root metaphors that Bowers find inhibits Western culture's development of ecological sustainable education are mechanism, anthropocentrism, dualism, individualism and progressivism. Bowers was inspired by the biologist and anthropologist, Gregory Bateson, who according to his daughter, Nora Bateson, in the film *An*

Ecology of Mind (2010), claimed: "The major problems in the world result from the difference between how *nature* works and the way *people think*."

Inspired by Bateson, Bowers set out on a crusade to crush the modern tenets in Western culture that dominate our educational system and to replace it with an alternative emergent, relational, co-dependent and recursive way of thinking. He found in Bateson many of his key arguments for this (Bowers, Jucker, Ishizawa Oba, & Rengifo, 2011, p. 5). Among them Bateson's, notion of "mind" and "nature" as a necessary unity, the theme of Bateson's second major book. That recursive ways of earlier misconceptions continue to frame current ways of thinking. A failure to recognize a difference between "map" and "territory," or as Alfred North Whitehead would say is to fall for the "fallacy of misplaced concreteness" when you mistake something abstract for what is concrete. Bateson's notion of "double bind" that prevents us from recognizing how supposed "progressive solutions" are deepening the crisis. Finally, Bowers found inspiration in Bateson for proposing educational reforms that challenges our taken-for-granted cultural assumptions inherited from the past.

Bateson's concept of "mind" is important in the context of this chapter. He speaks of both a smaller mind and a larger Mind (with capital M), not unlike how Naess speaks of the self and the Self. The larger Mind, for Bateson, appears in various connecting patterns as *Metapatterns*, i.e., "pattern which connects" (Bateson & Bateson, 1987, p. 199). The smaller mind is not some mental capacity inside our head but mental processes extend to the information exchanges that are going on in between us as we relate to our environment. "The mental characteristics of the system are immanent, not in some part, but in the system as a whole" (Bateson & Bateson, 1987, p. 32). All living beings, including the amoeba have mental characteristics. The problem with the ecological crisis, according to Bateson, is that we separate mind from the material structure in which it is immanent. When we try to think from the outside, we embark on a great error. Bateson says, "epistemological error is all right, it's fine, up to the point at which you create around yourself a universe in which that error becomes immanent in monstrous changes of the universe that you have created and now try to live in" (Bateson, 1972, p. 485). Inspired by Bateson, Bowers saw the need to critically question and revise the mental maps in our modern Western culture that are transmitted through education (Bowers et al., 2011, p. 28). In *Responsive Teaching* (1990), he says that a Batesonian alternative "starts from the premise that we must expand our understanding of mind outward to the point we recognize *self* as part of information exchange processes that constitute the ecology of which we are a part" (emphasis added, Bowers & Flinders, 1990, p. 23). This is close to Naess' concept of the larger Self, but just as Bateson speaks of a smaller mind within the larger Mind, Naess holds that the "individual self" do not vanish in the relational Self.

Of the many contributions Bowers proposed to educational reforms, which are too many to mention, I will settle on two: mindful conservatism and responsive teaching. First, Bowers seeks to "rectify the use of the words 'tradition' and 'conservatism' as a necessary first step in rectifying our relationships with each other and the environment" (2003, p. 121). He says, "mindful conservatism requires owning the word in a way where one takes responsibility for determining what is worth conserving, as well as responsibility for the expressions of conservatism that contribute to the many problems we face within our communities and with our cultures" (2003, p. 122). This is done, in part, by "revitalizing the commons" (Bowers, 2006), or focus on "the community and intergenerationally centered traditions of mutual support – that is, the traditions that represent alternatives to the commodified relationships, skills, and knowledge being promoted by the spread of corporate culture" (2003, p. 122). We need to be mindful to which traditions we need to conserve, which we need to discard and which we need to revise. The natural and cultural commons are, according to Bowers, under siege by corporate global culture. Bioregionalism is an important alternative, he argues, that can "regenerate the non-commoditized skills, knowledge and relationships that enables individuals, families, and communities to be more self-reliant and thus a smaller ecological impact" (Bowers, 2002, p. 14).

In Norway, the *Act of Common Access* ("*Allemannsretten*") can be understood as a cultural common. The act gives the public a right to roam freely in nature and camp anywhere within a reasonable distance to private houses and grazed land. This right is challenged by privatization and commodification of access to nature, e.g., through paid trails or enclosures. Other examples include certain skills in *friluftsliv*, such as skimaking, that are no longer freely accessible. Revitalizing this could reconnect us both to sustainable use of nature and thereby help to regenerate nature-friendly, bioregional communities and cultures.

The notion of *responsive teaching* is borrowed from the title of Bowers and Flinders (1990) book which defines "responsive" as "to be aware of and capable of responding in educationally constructive ways to the ways in which cultural patterns influence the behavioral and mental ecology" (1990, p. xi). Mindful teaching is both conservative and responsive in the sense that Bowers (and Flinders) propose. It helps decode taken for granted cultural patterns of communication that teachers use as significant others, which frame students' way of thinking and acting.

A Deep Ecocultural Approach

Bowers critique of Naess in the *Trumpeter* in 1993 was also directed at the Australian ecophilosopher, Warwick Fox, who at that time introduced the concept of "Transpersonal Ecology" as an alternative to "deep ecology," something Naess also was skeptical to, since Fox could be interpreted as less interested in the smaller self than the evolving larger Self. Naess was convinced that it is the other way around that the smaller self comes forth through the larger Self. Bowers criticizes on his part Naess for advocating a rationalist approach that emphasizes the individual and undermines culture. "The rational approach [in Naess] continues to reinforce cultural patterns that frame culture out of the picture" (1993b, p. 3). Naess' rational arguments, says Bowers, favors literacy-based discourse, which continues a liberalist, de-contextualized way of thinking that privileges detached body of knowledge and ignores culture. The most powerful leverage points for effecting change, according to Bowers, are at the level of culture.

Naess responds first to the critique of rationalism, individualism, and then of culture. He argues that his form of rationality is not Cartesian but rather influenced by the normative systems and values-oriented rationality of Spinoza. To be rational does not mean to take you abstractly out of the world, but places you concretely within it, as Spinoza does. When it comes to his focus of the individual, Naess defends this. He says we should not underestimate our potentials as individuals, "My experience in traditional societies has led me to believe that the range of personalities and range of choices of lifestyles may be quite considerable there" (Naess, 1993, p. 4). He continues, "I believe sincerely in the vast strength and importance of historical, social and political determinants of personal decisions" (Naess, 1993, p. 5). Naess then gives an account on how he understands the importance of cultural change but keep using the term "culture" in the anthropological sense, not in the deeper sense that Bowers advocates. Naess ends by stretching out a hand for collaborating on the necessary changes. He may not fully understand where Bowers comes from, but he recognizes it has an important place in the deep ecology movement. At the same time, he criticizes Bowers for not being clear on the role of the individual in cultural change.

Naess argues that it is easy from the outside to oversimply traditional culture as something collectively coherent, thereby overlooking how traditional cultures change. For Naess, traditional cultures change in part through individual differences. Naess says we should critique traditional cultures for their taken for granted beliefs and practices, e.g., in Balinese cockfight or the Spanish bullfights. Not for its social representation, which may in itself be problematic, but for its treatment of animals. Bowers is aware of the dangers in romanticizing traditional cultures, but he does not elaborate much on how traditional cultures change from within.

Naess wrote in the 1974 about "ecocultures" within the deep ecology movement (Haukeland & Naess, 2008, p. 240). He describes ecocultures as cultures that is characterized by minimal pollution, minimal resource exploitation, population control, self-reliance and resilience, appropriate technology adapted to the ecosystem the culture belong to, lifestyles of joy and realization of intrinsic values. The educational system within ecocultures, he says, should promote a high degree of sensitivity and valuing of goods that are plenty for all. It should further-more train students in non-violent conflict management, full participa-tion and competence distribution. There is a link between what Naess calls "ecocultures" and the ecocultural approach that Bowers promotes, but Bowers go deeper into the culture-language-thought connections.

Mindful Teaching in *Friluftsliv* Education

A deep ecocultural approach combines Naess' and Bowers' approaches. It is critical, constructive, recursive, and responsive. Mindful teachers in outdoor life (*friluftsliv*) education need to be attentive to their role as primary socializing agents. They should provide students with commu-nicative competence and deep ecocultural literacy. This is inspired by Bowers view on cultural literacy, which "designates not only the ability to speak the language and use it according to the norms of the cul-tural group, but also the ability to participate in (even to initiate) the public discourse about problematic aspects of social life...to bring out the political potential of language, highlighted by the mainstream cul-tural ideal of participatory decision making" (1993a, p. 180). Bowers expanded his ideas on cultural literacy to ecological literacy, referring both to David Orr's book *Ecological Literacy* (1992) and The Elmwood Institute's and Fritjof Capra's "Ecoliteracy program" (Clark, 1993). Capra (1996) sees "deep ecology" as the new, emerging paradigm and found also much in Bowers views. Bowers (1995) thought such initiatives needed a clearer focus on the language-thought-culture connection. He argues, "Ecological literacy, which too often is associated only with activ-ities that occur in schools, can now be broadened to include an aware-ness of how the assumptions, values, technologies, and categories of thinking of a culture influence how humans relate to the environment" (Bowers, 1996, p. 5). Naess in his own way concludes in his ecosophical remarks on *friluftsliv*: "With increasing understanding, increasing sen-sitivity toward internal relations, humans can live with moderate mate-rial means and reach a fabulous richness of ends" (Naess, 1989, p. 183). This presupposes a deep ecocultural consciousness and practice among teachers in outdoor life (*friluftsliv*) education, which is hard to do. We need more research and discussions on how to do so.

Steps to an ecology of mindful teaching in outdoor life (*friluftsliv*) education is directed toward the pedagogical training of outdoor life (*friluftsliv*) educators, so to become critically mindful and attentive to differences, patterns, relationships and intra-connections and contexts in teaching, learning and curriculum development. When educators help frame the learning process in terms of ecocultures, students are more likely to develop a greater sensitivity and attentiveness to the inter-connections between cultural and natural systems in the bioregion. Mindful outdoor life (*friluftsliv*) educators can teach about the habitat, natural and cultural commons, looking into traditional way of living, re-skilling, and symbiotic relationships. When nature is described in curriculum as a "learning arena" or "recreational arena," educators fail to see how it is our "home." In so doing, it makes students more alien to what it means to be at home in both nature and culture.

Mindful teaching in outdoor life (*friluftsliv*) education will benefit from combining Bowers (1984, pp. 86–95) principles for guiding curriculum development and Naess (1989, pp. 179–180) ethical guidelines for advocating *friluftsliv* in four ways. First, mindful teaching should utilize students' phenomenological culture, which gets at the aspects of the student's life world and natural attitude, teaching *friluftsliv* as a way of life and with a basic respect for life's intrinsic values. Second, mindful teaching should use historical perspectives to *de*objectify and *de*code taken for granted conceptual and interpretive frameworks, e.g., how objective knowledge undermines other ways of knowing grounded in the practice of *friluftsliv* and the process of identification. Third, mindful teaching incorporates cross-cultural perspectives, which includes the "more-than-human world" (Abram, 1996) and what Naess (1979) calls "mixed communities." It emphasizes dwelling more on together-ness and less on instrumentality. Fourth, mindful teaching in outdoor life (*friluftsliv*) education needs to be attentive to the promotion of deep ecocultural literacy, which includes minimal strain upon the natural systems.

Concluding Thoughts

A deep ecocultural approach to mindful teaching is relevant for promoting reforms specific to outdoor life (*friluftsliv*) education and more generally to curriculum development that directly addresses a sustainable living. The current Corona-crisis (starting Spring 2020) has made people more apt to reconnect to what is local but we should not lose sight of the global challenges of poverty, deteriorating ecosystems, loss of species, and climate change. A deep ecocultural approach to outdoor life (*friluftsliv*) education is, as I see it, *glocal* in scope, i.e., it needs to connect the global in the local and the local in the global. It should furthermore take a critical look at what is gained and what is lost in the transition from traditional to modern cultures. In this pursuit I have learned immensely from both Bowers and Naess. In their unique ways, they were curious, friendly, supportive, patient and inspiring to us, stumbling and fumbling students, but also demanding and challenging. While I went my own way, as they encouraged, I aspire to combine the same academic rigor and personal commitment they taught me. Chet

Bowers and Arne Naess, brilliant intellectuals and philosophers in the true sense of the word, as seekers of wisdom.

References

Abram, D. (1996). *The spell of the sensuous: Perception and language in a more-than-human world* (1st ed.). New York, NY: Pantheon Books.

Bateson, G. (1972). *Steps to an ecology of mind; collected essays in anthropology, psychiatry, evolution, and epistemology.* Toronto, Canada: Chandler Pub. Co.

Bateson, G., & Bateson, M. C. (1987). *Angels fear: Towards an epistemology of the sacred.* New York, NY: Macmillan.

Bowers, C. A. (1984). *The promise of theory: Education and the politics of cultural change.* New York, NY: Longman.

Bowers, C. A. (1993a). *Education, cultural myths, and the ecological crisis: Toward deep changes.* New York, NY: State University of New York Press.

Bowers, C. A. (1993b). Some questions about the theoretical foundations of W. Fox's transpersonal ecology and Arne Naess' Ecosophy T. *Trumpeter, 10*(3), 2–16.

Bowers, C. A. (1995). *Educating for an ecologically sustainable culture: Rethinking moral education, creativity, intelligence, and other modern orthodoxies.* New York, NY: State University of New York Press.

Bowers, C. A. (1996). The cultural dimensions of ecological literacy. *The Journal of Environmental Education, 27*(2), 5–10.

Bowers, C. A. (2002). Toward an eco-justice pedagogy. *Environmental Education Research, 8*(1), 21–34. https://doi: 10.1080/13504620120109628.

Bowers, C. A. (2003). *Mindful conservatism: Rethinking the ideological and educational basis of an ecologically sustainable future.* Lanham, MD: Rowman & Littlefield Publishers.

Bowers, C. A. (2006). *Revitalizing the commons: Cultural and educational sites of resistance and affirmation.* Lanham, MD: Lexington Books.

Bowers, C. A., & Flinders, D. J. (1990). *Responsive teaching: An ecological approach to classroom patterns of language, culture, and thought.* New York, NY: Teachers College Press.

Bowers, C. A., Jucker, R., Ishizawa Oba, J., & Rengifo, G. (2011). *Perspectives on the ideas of Gregory Bateson, ecological intelligence, and educational reforms.* Eugene, OR: Eco-Justice Press.

Breivik, G. (1978). To tradisjoner innen friluftslivet. In *Friluftsliv fra Ffridtjof Nansen til våre dager: Et utvalg.* Oslo, Norway: Universitetsforlaget.

Capra, F. (1996). *The web of life: A new scientific understanding of living systems* (1st Anchor Books ed.). New York, NY: Anchor Books.

Clark, E. (1993). How do we design an ecoliteracy curriculum? In E. Clark et al. (Eds.), *Guide to ecoliteracy: A new context for school re-structuring.* https://www.ecoliteracy.org/board.

Dahle, B. (1995). *Nature: True home of nature.* Oslo, Norway: Norges Idrettshøgskole.

Directorate of Education. (2017). Core curriculum kindergartens. Governmental documents. https://www.udir.no/laring-og-trivsel/rammeplan/

Directorate of Education. (2018). General curriculum compulsary education. Ministry of Education. https://www.udir.no/lk20/overordnet-del/

Fredman, P., Stenseke, M., Sandell, K., & Mossing, A. (2013). *Friluftsliv i förän-dring [Changing Friluftsliv]*. Final report, no. 6547 (1–336).

Gurholt, K. P. (2014). Joy of nature, friluftsliv education and self: Combining narrative and cultural-ecological approaches to environmental sustainability. *Journal of Adventure Education & Outdoor Learning, 14*(3), 233–246. https://doi: 10.1080/14729679.2014.948802.

Gurholt, K. P., & Haukeland, P. I. (2019). Friluftsliv and the Nordic model: Passions and paradoxes. In M. B. Tin, F. Telseth, J. O. Tangen, & R. Giulianotti (Eds.), *The Nordic model and physical culture* (pp. 165–181). London, UK: Routledge.

Haukeland, P. I. (2018). *Rikt friluftsliv med enkle midler*. Oslo, Norway: Stiftelsen NakuHel Norge Universitetet i Oslo, Institutt for helse og samfunn.

Haukeland, P. I., & Naess, A. (2008). *Dyp glede: Med Arne Nnæss inn i dypøkologien*. Oslo, Norway: Flux publisher.

Henderson, B., & Vikander, N. (2007). *Nature first: Outdoor life the Friluftsliv way*. Toronto, ON: Dundurn.

Hille, J., Aall, C., & Klepp, I. G. (2007). *Miljøbelastniner fra norsk fritidsforbruk: en kartlegging* (01). https://www.vestforsk.no/nn/publication/miljobelastninger-fra-norsk-fritidsforbruk-en-kartlegging.

Leirhaug, P. E., Haukeland, P. I., & Faarlund, N. (2019). «Friluftslivsvegledning som verdidannende læring i møte med fri natur». In L. Hallandvik & J. Høyem (Eds.), *Friluftslivspedagogikk* (pp. 15–32). Oslo, Norway: Cappelen Damm akademisk.

Ministry of Climate and the Environment (2016). *Friluftsliv: Natur som kilde til helse og livskvalitet*, White Paper no. 18 (2015–2016). https://www.regjeringen.no/no/dokumenter/meld.-st.-18-20152016/id2479100/.

Naess, A. (1979). Self-realization in mixed communities of humans, bears, sheep and wolves. *Inquiry, 22*(1–4), 231–241.

Naess, A. (1993). How should supporters of the deep ecology movement behave in order to affect society and culture. *Trumpeter: Journal of Ecosophy*. http://trumpeter.athabascau.ca/index.php/trumpet.

Naess, A. (1995). The Norwegian roots of deep ecology. In B. Dahle (Ed.), *Nature: The true home of culture* (pp. 15–18). Oslo, Norway: Norwegian College of Sports.

Naess, A. (1989). *Ecology, community, and lifestyle: Outline of an ecosophy* (D. Rothenberg, Trans.). Boston, MA: Cambridge University Press.

Naess, A., & Haukeland, P. I. (2002). *Life's philosophy: Reason & feeling in a deeper world*. Athens, GA: University of Georgia Press.

Orr, D. W. (1992). *Ecological literacy: Education and the transition to a postmodern world*. New York, NY: State University of New York Press.

Reed, P., & Rothenberg, D. (1993). *Wisdom in the open air: The Norwegian roots of deep ecology*. Minnesota, MN: University of Minnesota Press.

Sandell, K., & Sörlin, S. (2000). *Friluftshistoria: från "härdande friluftslif" till eko-turism och miljöpedagogik: teman i det svenska friluftslivets historia*. Stockholm, Sweden: Carlsson.

Veblen, T. (1965). *The theory of the leisure class1899*. New York, NY: A. M. Kelley Bookseller.

Part II

Curriculum of the Commons

The Industrial Revolution changed the community-centered craft guild systems of the medieval era in ways that reduced the worker to a wage-earner performing repetitive tasks required by the design of the machine. It also disrupted the rhyme of the community as well as the patterns of mutual support. The digital and global phase of this Industrial Revolution is now making not only the skills but the workers themselves redundant... (p. 108).

On the cultural margins of industrialized cultures, as well as within many non-Western cultures, the intergenerational knowledge and skills that sustained the cultural commons are being increasingly recognized as the only viable alternative to the fossil fuel and consumer-dependent culture ... These local, ecologically sustainable lifestyles can be supported by a fuller engagement in the cultural commons that exist in every community... (p. 104).

Cultural alternatives to the individually centered industrial/consumer-dependent lifestyle (now recognized as a major contributor to climate change) continue to exist in every community and in every culture. Indeed, they have existed since the first humans shared knowledge of where to obtain food, how to prepare it, how to bury their dead, how to develop technologies, and how to use them. That is, the knowledge, skills, and mentoring that were not monetized and thus shared intergenerationally among the first humans have been carried forward in culture-guiding mythopoetic narratives. These stories have been continually revised to fit the diversity of environmental challenges and changes in every culture... (p. 113–114).

Regardless of social status and level of formal education, everyone today relies upon the cultural commons of their family, ethnic group, and the larger shared culture. These intergenerational traditions of knowledge and skill are largely taken for granted and thus not recognized as an alternative form of wealth that exists outside the money economy. Increasingly they are being recognized as community-centered lifestyle alternatives that: (1) Have a smaller ecological footprint and (2) represent approaches to work where the individual's craft knowledge and skill lead to crating what is useful to the community rather than being driven by machine technologies to produce for the mass market. The cultural commons are recognized as absolutely essential in many cultures around the world. They continue to carry forward, in the face of the West's colonizing pressures, traditions responsive to the uniqueness of their bioregions and guiding mythopoetic narratives... (p. 114).

If our educational systems, media, and elites such as philosophers, social theorists, and even liberal arts graduates had depended less on technologies that promote abstract thinking and more on a balanced approach to learning that includes cultural awareness, there might be wider recognition of the importance of the cultural commons. The cultural commons, which everyone relies upon but largely takes for granted, includes the shared language, recipes for preparing food and preserving food, narratives, different creative arts, and linguistic patterns of meta-communication. We even rely upon those aspects of the cultural commons that reproduce the prejudices and misconceptions of earlier generations—in the same way that English speakers take for granted the subject/verb/object pattern of organizing and expressing their thoughts. And these thoughts largely reproduce the metaphorical patterns of thinking of earlier generations (p. 114). The point to be taken here is that, the taken-for-granted cultural commons need to be made explicit and critically understood as alternatives to the dominant culture—a culture now driven by techno-utopians whose proposals would: (1) Replace humans with machines in the workplace, (2) promote non-democratic decision-making, and (3) reduce human lives to data that will be exploited by corporations, hackers, and surveillance-addicted governments ... (p. 115).

Perhaps if the educational system had not relegated face-to-face and largely non-monetized intergenerational forms of knowledge and skills carried on in every community to such a lower status, then people would be able to acknowledge that expansion of the industrial/ consumer approach to progress will only accelerate further destruction of the environment. Awareness of climate change without any sense of the existing cultural alternatives leaves us in the same confused state of powerlessness as our leading politicians who are in denial about the relationship between consumer-dependent industrial culture and environmental degradation ... (p. 115).

Giving close attention to the otherwise taken-for-granted cultural ways of thinking and behavioral patterns, which are also part of the most broadly shared cultural commons, brings into focus other intergenerational traditions that need to be carried forward. These include such gains in social justice as forcing King John to sign the Magna Carta; or passing legislation to: (1) Prohibit the exploitation of child labor, (2) enable women to vote and escape previous cultural restrictions, and (3) allow for some environmental protections, and so forth. All of these social justice gains resulted when people became, so to speak, their own ethnographers. They paid attention to where and how they were being restricted by older prejudices; when and how they were being exploited by economic and political interests outside their own... (p. 115).

A "perfect storm" of destructive forces is coming together: (1) Chemical changes in the world's oceans, (2) rising temperatures, (3) loss of species and habitats, (4) increasing numbers of people facing food and water shortages, (5) the digitizing and globalization of industrial culture, with an accompanying loss of local knowledge and skills, (6) the increasing displacement of people by digital technologies, (7) impending perpetual armed struggle against colonization and increasingly scarce resources, (8) greater connectivity reducing every aspect of daily life to data, and (9) increasing control of the political process by ideologues and the super-rich. In the face of all of this, it is critically important to acknowledge resistance: People worldwide are recovering the importance of the cultural commons—and are finding ways to participate … (p. 115–116).

Unfortunately, these people represent only a small minority within Western cultures, and many indigenous cultures struggle to renew their cultural commons intergenerationally, as their youth are seduced by the commercialization and digital technologies of the West. Whether the diversity of the world's cultural commons will survive beyond the end of this century is problematic but they represent the best hope. They also represent community zones of safety from the surveillance technologies of hackers, corporations, and governments. Face-to-face relationships, local barter, and mutual exchange economies, and face-to-face decision-making about the needs of the community (i.e., local democracy) do not leave an electronic footprint that can be turned into the data profiles now used to invade people's privacy… (p. 116).

Why use the phrase "cultural commons" rather than "community?" The main reason is that the word, "community" is too general. It is too open to interpretations that lack a key part of the vocabulary associated with the phrase "cultural commons." What the commonplace use of the word "community" lacks is the special sense given to the cultural commons by another word, "enclosure." That is, the concept of the cultural commons includes the possible threat of enclosure, which is the process by which something this is shared in common is turned into something that is privately owned, then monetized and integrated into the industrial/market economy… (p. 116)

The vocabulary of the cultural commons is associated with the enclosure movement that swept through England during the Industrial Revolution, a time when the move to ownership of common lands benefitted wealthy landowners at the expense of those who previously had free use of common areas for crops and grazing. Thus, today's use of the phrase "cultural commons" has often been misinterpreted as an appeal to return to the pre-industrial lifestyle of earlier centuries. This misconception would not exist if people were explicitly aware of the cultural commons carried forward in their families, communities, and even within the dominant culture … (p. 116–117).

There is … a need to make explicit how the cultural commons are being enclosed. People born into a world that is saturated with commercials, with an overwhelming number of choices, with constant promises of how buying goods and services extolled by beautiful experts will add to even more happiness and success. All of this is so taken for granted that public spaces not exploited by the sounds and images of commercial culture become the oddities. We assume that assaults on consciousness and self-identity are a normal part of daily life. If this is all one knows, it will feel more like home and less demanding … (p 119).

Ethnographic mapping needs to be done to show how different technologies may be empowering, but also work together to undermine the very existence of the cultural commons. As noted earlier, an over-reliance upon print-based cultural storage and thinking promotes the Western cultural bias toward abstraction, which fosters the illusion that rational, even critical thought is supposedly free of cultural influences. This reinforces the misconception that an individual has a unique, unmediated experience in the world, and that giving close attention to cultural patterns that connect is irrelevant ... (p. 119).

The failure to do the ethnographic work of making these patterns explicit is compounded by the failure of classroom teachers and university professors to provide the language and theory frameworks that would enable students to recognize aspects of their taken-for-granted worlds that are sources of empowerment and community. When classroom teachers and professors *do* bring cultural commons practices into focus, it is usually to help students recognize patterns of discrimination and exploitation, of which there are many. This is vitally important, but it is generally framed in terms of an ideology that reinforces the idea that the primary purpose of critical thinking is to overturn all traditions that stand in the way of achieving social justice and thus progress. This ideology does not consider critical attention to ideas that would reduce our dependence upon a money economy and have a smaller adverse ecological impact, nor does it consider the cultural commons as a source of skills and supporting relationships... (p. 121).

*Excerpt from:

Bowers, C. A. (2016). *Reforming higher education in an era of ecological crisis and growing digital insecurity*. Anoka, MN: Process Century Press.

6 On Traditions and the Commons

A Material Feminist Analysis

Audrey M. Dentith

For more than 25 years, feminist thought has informed nearly all of my scholarly work but no scholar has influenced my thinking more than Chet Bowers. As a mentor, scholar and teacher, he embodied a feminist ethos and his thinking about justice, indigenous knowledge, culture, and ecojustice education have greatly influenced me. I met Chet for the first time in 2010 when he came to the University of Texas San Antonio to do a keynote address. I had read some of his work prior to this visit but did not know him personally. At the time, I was involved with an interdisciplinary Hispanic cultural project in teacher education and his work on the cultural commons and indigenous knowledge resonated with and strongly influenced that work. Our ensuing friendship along with his generous ability to mentor and connect with me and my students through subsequent years offered many ways for me to think more deeply about my feminist politics amid the deepening environmental crisis we now face.

In a conversation I had with him in 2012, he encouraged me to think about the connection of the cultural commons with feminist theories and to write about it. The purpose of this chapter, then, is to fulfill that challenge as I consider the relationship of Bower's conception of the cultural commons to concepts within material feminist thought. To do so, I hope to bridge the ecological, the biological/physical, and the cultural/social. To make this connection, I begin with his ecojustice framework and the definitions of the cultural commons. I, then, examine the evolution of Bower's thinking from the work of Shils' (1981) on tradition. I, then, turn to an examination of essential elements of feminist materialist thought. Finally, I bridge Bower's work on the cultural commons to the contemporary work of feminist materialism, demonstrated by definitions, tenets and linkages to both bodies of thoughts.

Ecojustice Education

Bowers urged us to focus on the cultural roots of the ecological crisis and engage with others to understand the importance of acknowledging patterns of language and culture in living lives that embrace more sustainable practices. For Bowers, the ecological crisis is a cultural crisis. Language and culture must be understood and interrogated if we are to overcome the many challenges wrought on by humans' degradation of the environment. Bowers called his approach, *Ecojustice Education,* and it formed the foundation for his life's work, which, at its core, relies on the fundamental belief that we are all engaged in a complex system that is interdependent and reliant on the health and well-being of all others, human and non-human (Bowers, 2001).

Bowers identified at least six tenets of an Ecojustice Education, in all, but the principle that most engages me in this chapter is his call to preserve the natural and cultural commons. The natural commons (the air, land, water, vegetation, and species) are being consistently eroded by over-population, pollution, problems of waste disposal, climate change, ocean acidity, etc. and threatened by forces of industrialization and privatization (Bowers, 2001). Bowers strongly advocated for the protection and revitalization of the "cultural commons." He maintained that a revitalization of the culture commons would promote lifestyles that resist the expansion of the industrial/consumer dependent lifestyles. He thought that a life that embraces the cultural commons would lead to ways of living that respect the right of natural systems to renew and not further degrade the environment. The practices would also support the cultivation of local democracies and lead to more community-centered alternatives to the "deskilled individual lifestyle that is increasingly dependent upon consumerism" (Bowers, 2018, p. 156).

The cultural commons are the self-sustaining, largely non-monetized and face-to-face practices, skills and knowledge that exist in every community and have endured for thousands of years. These include indigenous languages, narratives, craft knowledge, games, medicinal practices, art, dance, the traditions of food preparation, storytelling, and mentoring, and so forth (Bowers, 2012). The revitalization of the cultural commons, is a radical departure from consumer-laden prevailing practices. Bower believed that the cultivation of the cultural commons will ensure survival for billions of people in the years ahead as well as slowing down of the plundering of the world's natural resources (Bowers, 2001, 2006). In many cases, the cultural commons in a given community are the basis for local economies, and their accompanying systems of mutual support sustain their survival and means of intrinsic wealth and prosperity. This support is in contrast to the liberal market economy which is most often driven by an incessant demand for a constant stream of new products. He claims that without a concerted effort to revitalize the cultural commons, we are in danger of being complacent in their "enclosure." Enclosure is the further promotion of market ideologies and the privatization of the natural commons. The cultural commons counter the prevailing forces of enclosure. They are grounded in moral values that are shaped by positive, engaged experiences shared solely in the physical presence of humans (Bowers, 2011a, 2016, 2018).

While the nature of the cultural and environmental commons are as diverse as the world's cultures and bioregions, what they have in common is that they represent what has not yet been monetized and brought into the industrial approach to markets. In effect, the cultural commons represent cites of resistance to the spread of a money and consumer dependent lifestyle—and thus to the spread of a world monoculture and to the further spread of poverty for those who lack the means to participate in a money economy (available at: http://cabowers. net/CAbookarticle.php).

Tradition and the Cultural Commons

Bowers was often the target of critique for his advocacy of a revitalization of the cultural commons. Conversely, he cautioned against the romanization of the cultural commons reflecting on the fact that many would not recognize the wealth of the commons since they do not always fit with current norms of social and ecological justice (2012). His numerous critics claimed it was a throw-back to the past, a romantic call to bring back practices of the past and an unrealistic effort that could not ever be fully accomplished (Bowers, 2006, 2016). This greatly simplified the scholarship and theory that undergirded his call for revitalization. The answer for Bowers was not to "go back" but, instead to learn from all the traditions of diverse culture in order to imagine how these practices still exist in our everyday life. This means a deliberate effort to recognize the sustained cultural practices that do not lead to further destruction of the environment and that promote a move from a money-dependent method of exchange and a life less dependent on outside forces and consumptive practices.

Bowers' initial thinking about the revitalization of the cultural commons came from his study of Thomas Shils' (1981) which Bowers claimed as one of the most important reads in his scholarly life. Shils' views on the importance of acknowledging, how the past needs to inform the present is evident in this assertion that very few argue for the revival of the past. Once a situation is brought to light, there is a presumption that it ought to be changed as a measure of improvement. One may find fascination with past practices, but the value of beliefs and patterns of the past are often diminished. Progress is equated with the need to change, to replace, to discard. Change is aligned with progress. Scientific knowledge is thought to be the antithesis of traditional knowledge and, as we've come to realize, the cause of much of the world's ecological destruction. Scientific procedure and rational criticism is contrasted with knowledge of elders. Social scientists often dismiss tradition as something to be brushed away.

So, tradition is not to say that things stay the same. Traditions do undergo changes over time, but what makes a tradition is that *what* are thought to be essential elements remain recognizable by an external observer as holding similarities at successive steps or acts of transmission. Thus, tradition is not a romanticized return to the past, but an evolution of continuity and connectedness over time on the part of the observer (Shils, 1981).

Shil's assertion can be understood in many facets of life. When I am bent over the vegetable boxes in my own garden, I am reminded of my grandmother whose large garden provided fresh food for all of us over the summer and early Fall. In her house dress with the big apron, she'd gather the tomatoes, peppers, and zucchini by folding her apron back to meet her body. I find myself immersed with my memories of her when I'm picking from my own garden, bringing back happy times of food traditions: Large pizzas topped with fresh tomatoes and basil, zucchini flowers sautéed with olive oil and fried peppers with sausage bits. Although my own gardening and cooking practices have changed somewhat over the years, the lessons learned from these traditions of gardening and cooking have informed my own practices for more than a half-century.

Moving beyond the Social/Cultural

For feminist materialists, tradition requires a strong consideration of biology and the physical realities in a call for an end to sexism and the denigration of those marginalized in the world. Cultural traditions, social, political evolve in much the same way as the physical world. Materials feminists regard Darwin's theory of evolution as important to understanding the evolution of change over time. Previously, ontology has long been considered to be static fixed, indifferent to history and change. Now material feminisms acknowledge realities beyond the immaterial and embrace temporal forces of endless change. Feminist theories should not be just abstract and relevant only to political processes. A new complex view of culture is significant (Grosz, 2007). On this point, Grosz (2005) urges:

> feminists to regard with some rigor and depth the usefulness and value of Darwin's theory of evolution on our conceptions of culture, social, political, and sexual life in order that we see these things as more complex, more open to questions of materiality and biological organization, more nuanced in terms of understanding both the internal and external constraints on behavior as well as the impetus to new and creative activities. (p. 14)

Feminism, she asserts, can learn much from the principles of Darwinism and theories of evolution to produce richer, more workable concepts of "nature, the body, time and transformation than those available only through the discourses of cultural and political theories, history and philosophy alone (p. 17)."

Materialists do not dismiss the recent gains made by prevalence on the cultural/social notions of a feminist politics; instead, they embrace the past and look to ways that biology can inform and push forward an over-reliance on the abstractness of the linguistic and the cultural/social.

Ecojustice Education and Ecofeminism

Bowers' work includes an ecofeminism perspective, particularly as it relates to his call for earth democracies and an acknowledgement of the ways that women and other marginalized people are more subject to environmental injustice due to their subjugated position in the world.

Ecofeminism borrows from and builds upon the knowledge of scholars in Marxist, liberal, radical, multicultural feminisms. Ecofeminists are centrally interested in the relationship of women and nature, moving beyond a focus on the "social isms of domination" (Warren, 1997, p. 4) to regard this new perspective. Ecofeminism is "distinct in its insistence that nonhuman nature and naturism are feminist issues" (Warren, 1997, p. 4). For ecofeminists, the "woman-nature connection" (Warren, 1997, p. 5) is foundational since women, particularly those from developing countries, are highly dependent on natural resources: Water, food and farming, wildlife, and trees and forests. Moreover, poor women and women of color across the world are often the most adversely affected by all forms of pollution, water shortages, and natural resource depletion. Ecological destruction is, according to Mies and Shiva (2014), a direct threat to millions of poor women across the world. Because of the roles they play in their families and communities, women are more susceptible to the ill effects of ecological degradation. Silvia Federici (2019) extends the work of ecofeminism and the commons by offering an extended Marxist critique of capitalism through the lives of women in the commons across the globe. Ecofeminism has also been characterized as a "romantic conception of both women and nature, the idea that women have special powers and capacities of nurturance, empathy and 'closeness to nature' which are unsharable by men" (Plumwood, 1993, p. 8). Ecofeminism is certainly compatible with Bower's work, but it is feminist materialism that may provide a more embodied explanation of the relationship between feminism and the commons.

Feminist Materialism

While it is beyond the scope of this essay to fully address the distinctions between ecofeminism, Marxist, radical feminism and material feminisms, a general understanding of feminist materialism will be offered.

Feminist materialism implies a movement that extends or reforms the linguistic or cultural turn to one that (re)considers the material. It is a "movement from epistemology to ontology" (Heckman, 2010, p. 68). Feminist materialism is "a concerted effort to define an alternative approach that brings the material back in" (Heckman, 2010, p. 4). Moreover, in this conception, there is not a privileging of the material or the cultural because all production depends on both to function. According to Grosz (2005), nature is not the polarized opposite of culture but rather an underlying condition. For feminist materialists, nature is a dynamic force within "fields of transformation and upheaval, rather than as a static entity, passive, worked over, transformed and dynamized only by culture" (p. 7).

Below I discuss two premises of feminist materialism, including: (1) The agency of all matter and an acknowledgement of the life-force of the physical, subhuman, and other-than human entities; and (2) the notion that there exists a blurring of boundaries between the physical and the cultural. In this configuration, things are not separate but emergent, shifting, and complex in their relationality, more than just interactive, entities are unable to be independent or self-contained from other matter (Barad, 2007; Tuana, 2008).

The Agency of Matter

A large black compost barrel sits at the edge of my garden. Every day we dump our organic waste into the black hole at the top of the contraption. It has a foul smell and when you open the cover; you are immediately aware of the noise that emanates from it. It's a vibrant ecosystem... meal worms, maggots, and flies are all gnawing on the various discarded food items. This is not a just a disposal site nor a simple pile of organic matter. It's alive in active, resistant, congealing, and turbulent activity. Nonhuman life has agency. It's possible to see even beyond the entangled mess to the oozing of black matter coming down from the holes in the side of the barrel, making its way to the garden where worms and other matter will feast or let the matter glide into the ground, altering the physical composition of the garden soil and eventually the food that grows within it and is consumed by humans. This activity is not passive or separate from the human experience

With all forms of bacteria, there is essential and real action taking place underneath. The back-staging and metabolizing work of microbes and bacteria create the possibilities for larger lives and actions. It is essential life-supporting work that is going on beneath and around us all the time, where quadrillions of worms and ants are at work creating the show that we, as humans, act out (Hird, 2012). All subhuman matter, then, has agency, or the ability to act or move in ways that affect other beings (Bennett, 2010; Plumwood, 2002; Snaza, Sonu, Truman, & Zaliwska, 2016).

This is fundamentally about the reworking of older materialist traditions that might address pressing ethical and political challenges in the world. A post-humanist conception of matter as lively or exhibiting agency and a re-engagement with both the material realities of everyday life and the broader geopolitical and socioeconomic structures is evident in the work of many contemporary feminist scholars. Feminist materialists see the place of embodied humans in a material world and in the ways that we produce, reproduce, and consume our material environment (Frost, 2011; Grosz, 2010).

The Blurring of the Boundaries among All Matter

When my daughter became a mother, she breastfed her babies. She joined many other contemporary mothers who have revitalized the practice of feeding babies with breastmilk. This effort is in contrast to the production and proliferation of baby formulas which began in the 1950s. Corporate manufacturers claimed, untruthfully, that factory-produced formulas were more nutritious and better for babies than breast milk. Maintaining the traditions of breastfeeding has been supported for more than 60 years through an international group, *La Leche League*, which began with six women in Chicago in 1952. These originators refuted the claims of the corporate promoters of their time and formed a breastfeeding club that reached out to women with scientific information on the nutritional value of breastfeeding and instructions for new mothers on how to breastfeed. Beyond distributing information, these women supported other women directly in their communities through meetings and informational gatherings in church basements and community centers. They supported one another through breastfeeding practices, claiming that not only does breastfeeding provide natural nutrition and important maternal antibodies to babies but it also fosters bonding and strengthens family relations. The international success of their work has spread to hundreds of countries across the globe. La Leche League has significantly contributed to the return and sustainment of the practice of breastfeeding among contemporary mothers (https://www.lllusa.org).

Here we see the blurring of the boundaries between the physical and the social/cultural world in this practice that revitalizes the cultural commons. Feminist materiality always regards the human body and the natural world as central to feminist theory and practice. The ontology of women and their embodied experiences are interwoven within a feminist-situated epistemology. Karen Barad's work on agential realism is instructive here. Rather than adhering to a binary between the material and the cultural, she describes the processes of the material and the cultural which interrogates the perceived but false boundaries between them. Agential realism emphasizes an ethics that reflects and respects complexity and the intra-dependency of all that is natural, corporeal, social, and cultural (Barad, 2007).

Part of understanding the importance of tradition relies on an engagement with the material experience of living our lives. Materialist feminists acknowledge the ways that constructivist and cultural epistemologies have disentangled woman from nature and embodied experience. Feminist scholarship in the cultural/linguistic turn have regarded nature as primarily a kind of obstacle against which there is a need to struggle, not work from or within it. Such feminist politics have viewed nature as inactive, unchanging and resistant to social, historical, and cultural transformations. Material feminism, on the other hand, extends the focus on nature, biology and raw materials from the onset. Natural forces were evident at the beginning but were disregarded or undervalued by the cultural turn feminism experienced in the last generation. "The others who are inhuman, subhuman and extra human forces, those that serve as the structure, law and representation and all other products of the human—need to be understood in terms of a continuity with rather than in opposition to the human" (Grosz, 2005, p. 5). Barad (2007) calls this "intra-activity" while Tuana's (2008) names this as "Vicious porosity."

Feminist materiality pushes to recognize the value of the material and the vestiges of past practices which are particularly significant to the experience of humans. The new material feminist stress that the material exchange between bodies, consumer objects, and substances serves as sites for political and ethical engagements and as sites for intervention. The politicized actions that wed traditions with corporeality demonstrate how forms of activism indicate that "human corporeality and human practices are immersed within environments and affected by particular, embedded encounters" (Alaimo, 2016, p. 28).

Young mothers' commitment to breastfeed their children is a feminist, corporeal practice that relies on traditional knowledge and practice. It is also a political statement that pushes back against prevailing corporate practices. For material feminists even the smallest, parochial ethical practice in the domestic sphere is also tied to political and economic predicaments including capitalism, labor issues, pollution, and class injustice, industrial agriculture practices, climate change, and the extinction of many species.

A corporeal mode of habitation can also be made visible in the transformation of humans who engage in activities that connect them to the natural world and to which they become part of the natural entity that once existed outside of themselves. Stacy Alaimo indicates that material feminisms make space for the "active, emergent significance of the materiality of the bodies, substances, and environments" and, thus, create new possibilities for alliances between environmentalists and feminists (Alaimo, 2016, p. 12).

Tradition in the New Materialities

The above examples illuminate the merging of concepts of tradition and the cultural commons with feminist materiality. Materiality further extends traditions and cultural practices by considering food practices as more than just about the activities of people.

The slow food movement is instructive here as well. Begun in 1986, the Slow Food Movement, in which the growing, cultivating, and pre-paring of food extended into discussions of food as tradition, temporal activities, and political issues. The Slow Food Movement opened up a politicization of food within an embrace of traditional food practices. It effectively acknowledges tradition in the ways Shils described such practices and in the ways that Bowers suggested in his call for the revi-talization of the cultural commons. Learning and engagement are achieved in face-to-face relationships that acknowledge change but stay tethered to those values and practices that promote healthy life and local democracies. The slow food movement, however, not only focuses on the time to prepare the food and the tradition of food preparation, it also extended opportunities to reflect on economic, labor, agriculture, and transportation issues that come before the arrival of food to mar-kets and, ultimately, to our tables (Bennett, 2010).

Bennett (2010) asserts that the assemblage of the slow food move-ment, for example, could be strengthened further if it broadened its notions of food beyond the activities of humans. The movement tends to perceive of food as simply the means to fortify humans which, in turn, perpetuates the notion that non-human entities are passive, only one side of an ontological split between humans and matter. When we acknowledge the agency of food, we are then also able to reflect and adjust our own experience of eating. If the slow food movement incor-porated a greater sense of the active vitality and agency of food, a deeper more material understanding of food, itself, can be realized. What was once considered inert matter comes to be considered as having agency, and is experienced as a lively force, ripe with the capacity for agency. All of this can then allow humans to regard food in ways that might bring about a new consciousness among people (Bennett, 2010). In material-ism and within an understanding of the agency of nonhuman life, we are in "unexpected collaborations and combinations, in hot compost piles" (Haraway, 2016, p. 4).

Tradition relies on the appreciation of the accomplishments and wisdoms of the past as practiced by some indigenous people and is premised on the "understanding that the commons (plants, animals, rivers, humans, etc.) are equal participants in the same spiritual uni-verse" (Bowers, 2006, p. 94). These practices of the past often inform the present in ways that benefit the human and non-human world. Bowers named these traditions (2016) as those that go beyond just historical knowledge. These traditions involve knowledge of other cultures, particularly those that recognize the emergent, relational, and co-dependent nature of the world. Wisdom traditions are in contrast to the modern, secular, and data-based new world order, which is partial, rational, abstract, and not cognizant of a larger dynamic context (p. 87). Wisdom traditions, most often evident in indigenous cultures, stand in contrast to the modern mindset and have the most relevance for learning to live less environmentally destructive lives. They regard the interdependencies of life as fundamental to their existence. In Western societies, the goal of acquiring knowledge is to achieve greater control over natural processes. Indigenous knowledge, on the other hand, is "governed by a deep sense of moral reciprocity and of the sacredness of natural processes" (p. 97). In these traditions, here is a continued acknowledgment of nature's bounty within an experience of kinship, an understanding the interconnected patterns of all of life and the awareness of the ways that humans need to adapt in order to sustain life, all of which are in sharp contrast to a world of surface, abstract representations so prevalent in our modern world (Bowers, 2006, 2016).

Materialists regard bodies, spaces, and activities to be the nuclei of activity and engagement of all entities. Materialists do rely upon and value cultural practices ground in tradition. Haraway suggests that it's contemporary indigenous people that make a sensible difference (p. 5) as they consider the agentic nature of the nonhuman, the subhuman in the underlying phenomena of growing food, its cultivation and consumption as the new politics of food. Raw materials are the inception, a focus on nature, biology, and the inhuman forces that now inform us. Like the Slow Food Movement and the picking of produce from my garden, the traditions of food are evident but are only considered in the context of human agency, not in the agentic potential of nonhuman. It is here that tradition and the cultural commons can converge with materialism.

Traditions develop from the desire to create something truer and better is alive in those who acquire and possess them. All accomplished patterns of the human mind, all patterns of belief, or modes of thinking, all achieved patterns of social relations are susceptible to becoming objects of transmission, capable of becoming a tradition. Intimate association, however, is required in the transmission of tradition. Tradition relies on memory and embodied experience (Shils, 1981). From this assertion, Bowers forged his ideas about the necessity of face-to-face interactions in the cultural commons. Moving this forward, and extending this assertion in the new materiality, the embodied experience is foundational to all matter, human and non-human life.

Material feminism, to extend this discussion, does not depend solely on concepts of culture, subjectivity and identity which are concepts that have been at the heart of a feminist project. They consider food practices, tradition, culture, and identity as well as the agency of and nature of food itself. For feminist materials, "nature is understood in terms of dynamic forces, fields of transformation and upheaval, rather than as static entity, passive, worked over, transformed and dynamized only by culture" (Grosz, 2005, p. 7).

Materiality and the Commons

Nature is dynamic and active, not passive and culture is a "continuation and elaboration of nature rather than an overcoming of it" (Grosz, 2005, p. 8). Grosz advocates for a materiality that refutes duality and retains ideality; however, she does not privilege ideality over materiality, but approaches them as entangled, connected and unable to be without the other. Ideality and materiality are the implicit conditions for the other. It is the interconnected entanglement that generates the possibility for the emergence of the arts and science, particularly in ways that allow people to address and enhance the real and move toward goals and ideals (Grosz, 2017). Fundamentally, for Bowers, the integration of the material enacted in the activities of the cultural commons leads to a similar space, a place that offers people ways they might embrace the material with the ideal. His work suggests a kind of awakening that weaves the past into the present while illuminating the possibilities for renewed material life practices through both physical and social activity (Bowers, 2006).

In this new and potentially exciting space, there is a deep interconnectivity and the impossibility of a single reality, and I agree with Bennett (2010) who believes that "ecological sensibility" (p. xi) is important to understand, as there does *not* exist "a smooth harmony of parts nor a diversity unified by a common spirit" (p. xi). Materiality is, instead, characterized by a "turbulent, immanent field of which various materialities collide, congeal, morph, evolve, and disintegrate" (p. xi). Matter is comprised of active, entangled, disruptive spaces within a web of interrelatedness. This connectedness means that one thing is connected to another and another all of which make up our physical as well as our social world (Bennett, 2010).

Materiality, the Commons, and Education

Bowers believed that learning, particularly, in higher education is too heavily reliant on a transfer of abstract knowledge, in a process that typically occurs in ways that largely ignore the material (Bowers, 2016). Early on, he declared that an ecojustice pedagogy is centered on understanding relationships within the larger spheres of community and the natural environment. A contrast to the prevalent abstraction of learning in most educational institutions particularly in higher education,

Bowers insisted that learning be ground in physicality, immersed within communities. He was critical of the abstract theoretical and historical knowledge and would agree with materialists who decry the lack of focus on those activities that examine the physicality of learning (Bowers, 2016). For Bowers, education for the commons was the intent and the future of education. The abstraction of learning is antithetical to commons education... learning is material, physical, ground in physicality (Bowers, 2016; Hickey-Moody et al., 2019).

One example of such learning is Chet's and my work on the revitalization of the commons in higher education. We created seminars and institutes that highlighted the cultural commons in university education. Our first *"The Academy for the Critical Inquiry of the Cultural Commons,"* held in San Antonio, Texas in 2012, was an excellent example of ways to integrate knowledge and experiences of the cultural commons into a contemporary university education. We brought together activists, scholars, students, restaurant owners, ecologists, and other community members as presenters, authors, and learners to imagine a reworking of the university in which it becomes a true hub of engaged dialogue around the most pressing issues of our modern world. In this five-day academy, activists, scholars, undergraduate and graduate students, faculty, and community members assembled to share their knowledge and experience of the cultural commons. Artists, crafters, food from organic and sustainable restaurants, an activist in water conservation, films, academic speeches, and a food canning demonstration were some of the activities of this exciting academy. Graduate and undergraduate students, all in teacher education, created projects that blended academic knowledge with the cultural commons and, for the first time, became acutely and expressively aware of the presence of the cultural commons in their everyday life.

The commons education consists of a curriculum that seeks creativity, physical activity, emanating from the wisdom traditions of the past as the context for learning. Creativity is, at once, a physical bodily experience that offers both a theoretical understanding as well as an engagement in the physical actions of creating and being in the present. In their essay on dance, feminism materialism, and the arts, Hickey-Moody et al. write about materiality and the curriculum, remind us that "creativity is activity that produces something new, such as an idea or a tangible output" (p. 101).

Learning in the material is most likely to occur in interdisciplinary arrangements using diverse pedagogies, all of which "give voice to a multiplicity of learner subjectivities" (Hickey-Moody et al., 2019, p. 101). The material curriculum of the commons addresses the lack of interdisciplinary work in schools. The inability of teachers to recognize the interconnectedness of the physical, biological with the culture and politics stems from their own education. Bowers maintained that universities and schools are silent and dismissive of the wisdom traditions so fundamental to our survival. He thought that schools in Western societies undermine the values and knowledge of relationally informed wisdom traditions. Western schooling most often regards humans as autonomous and mechanistic beings in an anthropocentric world that revolves around people. Computer-mediated thinking devalues wisdom traditions even as it is becoming more and more evident that Western culture is accelerating the global changes that threaten the very existence of humans (Bowers, 2016).

As educators, we can consider the possibilities brought on by material feminism and tradition in our work to revitalize the cultural commons in education. Bowers (2011b) advocated for a "commons" curriculum. He thought that ecologically problematic cultural assumptions should be part of our discussions and learning about the commons and could be an integral component in every area of the curriculum, rather than be treated as a separate subject. Doing so would garner awakenings and lead to lifestyles among people that help them to understand the interrelatedness of the living with the non-living. This would break the prevailing silence and encourage a focus away from a separation of the self from the earth, toward a reconstitution of the ordered world in which the connections between things creates new meanings (Grosz, 2017).

Teachers as Mediators

For Bowers, it is important that we see ourselves (scholars and teachers) as mediators who mediate between the forces of modernization and the self-renewing characteristics of the cultural and natural commons by a continuous assessment that makes explicit the patterns of the natural and cultural commons vs. the industrialized culture. He urged us to bring forward the silences and misconceptions of the past in order to reimagine a different future. In addition, he maintained that our embodied experiences would support systems of cooperation and moral reciprocity (Bowers, 2011b).

To be mediators, Bowers claims that we need to name the practices of the cultural commons. Without this explicit naming, others will likely assume these to be part of their taken-for-granted experiences and will, thus, not see the values of these practices, not only for promoting better ecological practices but also for recognizing civil liberties and freedoms brought on by engagement with the commons. He also believed that we should discuss the patterns of thinking and values that are taken for granted, to break the silence regarding the fact that much of our knowledge is garnered at a pre-consciousness level of awareness. Bowers urges us to compare and contrast experiences between the monetized and enclosed activities to those activities represented in the cultural commons (Bowers, 2011a, 2011b, 2018).

Conclusion

Feminist materialism seeks to find a continuousness of harmony rather than discordance. While it can be easy to see the connections between Bowers' work and the work of various ecofeminists as a natural fit, the work of materials feminists offers a more robust, complex, and potentially more politicized feminist explanation. Material feminisms are able to speculate on the relationship of culture, politics, tradition, and the physical or material realities around us. Materialist feminist scholars also critique globalization and argue for more than a material focus to include one that values culture, language, and discourse as well.

If we consider and continue to build on a feminist materiality, we can then consider all bodies as a site of learning even as we continue to raise questions about how diverse bodies do fit in those environments that have traditionally suspended the body altogether, such as the university.

Like Bowers, Grosz (2005) consider concepts of time as a foundation to our very existence. Temporality results in new concepts of nature, culture, subjectivity, and politics. They are explorations of how far we can push the present to generate the unknown—what is new, what might not have been (p. 10). At this particular moment, it is imperative that we consider the temporality of the material in order to uncover new concepts of culture, subjectivity, and politics to push the present to find the as yet unknown. Here is where Braidotti (2019) can also be instructive. In her lecture given at Harvard University, she describes herself as an eternal optimist. Bradiotti views the present as offering knowledge of the past along with the promise of the future. Feminist materialism, which considers a new social and physical space that is active and engaged, ground in the corporeal and evident in the embodied experience of all humans, offers a way to envision the future of the cultural commons.

Tradition, the commons, and feminist materiality intersect and share commonalities. Tradition is vitally important to social and political institutions, technology, science, literature, and religion. It is not a reveling in the past but a look at the continuity and relevance of the past as it plays out in the present. The commons is a site of engagement of tradition with the present. An embodiment of culture with physicality. Feminist materialism in about feminist politics and the corporeal. Together, tradition, the commons and materiality offer us a deeper look at the possibilities for the practices and politics of the present into the future as we grapple with the dilemmas evident in our world.

References

Alaimo, S. (2016). *Exposed: Environmental politics and pleasures in posthuman times.* Minneapolis, MN: University of Minnesota Press.

Barad, K. (2007). *Meeting the universe halfway*. Durham, NC: Duke University Press.

Bennett, J. (2010). *Vibrant matter: A political ecology of things*. Durham, NC and London: Duke University Press.

Bowers, C. A. (2001). *Educating for eco-justice and community*. Athens, GA: The University of Georgia Press.

Bowers, C. A. (2006). *Revitalizing the commons: Cultural and educational sites of resistance and affirmation*. London, UK: Rowman and Littlefield.

Bowers, C. A. (2011a). *University reform in an era of global warming*. Eugene, OR: Ecojustice Press.

Bowers, C. A. (2011b). *Educational reforms for the 21st century*. Eugene, OR: Ecojustice Press.

Bowers, C. A. (2012). *The way forward: Educational reforms that focus on the cultural commons and the linguistic roots of the ecological/cultural crisis*. Eugene, OR: Ecojustice Press.

Bowers, C. A. (2016). *Reforming higher education in an era of ecological crisis and growing digital insecurity*. Anoka, MN: Process Century Press.

Bowers, C. A. (2018). *Ideological, cultural, and linguistic roots of educational reforms to address the ecological crisis: The selected works of C. A. (Chet) Bowers*. New York: Routledge.

Braidotti, R. (2019). *Posthuman knowledge*. Lecture at Harvard University. https://www.youtube.com/watch?v=0CewnVzOg5w.

Federici, S. (2019). *Re-enchanting the world: Feminism and the politics of the commons*. Oakland, CA: PM Press.

Frost, S. (2011). The implications of the new materialisms for feminist epistemology. In H. E. Grasswick (Ed.), *Feminist epistemology and philosophy of science: Power in knowledge* (pp. 69–83). New York: Springer.

Grosz, E. (2005). *Time travels: Feminism, nature and power*. Durham, NC and London: Duke University Press.

Grosz, E. (2010). The untimeliness of feminist theory. *Nordic Journal of Feminist and Gender Research, 18*(1), 45–51.

Grosz, E. (2017). *The incorporeal: Ontology, ethics, and the limits of materialism*. New York: Columbia University Press.

Haraway, D. J. (2016). *Staying with the trouble: Making kin in the chthulucene*. Durham, NC and London: Duke University Press.

Heckman, S. (2010). *The material of knowledge: Feminist disclosures*. Indiana: Indiana University Press.

Hickey-Moody, A., Palmer, H., Ringrose, E. S., Warfield, K., & Zarabadi, S. (2019). Diffractive pedagogies: Dancing across new materialist imaginaries. In E. S. Ringrose, K. Warfield, & S. Zarabadi (Eds.), *Feminist posthumanisms, new materialisms and education* (pp. 94–110). New York: Routledge.

Hird, M. (2012). Knowing waste: Towards an inhuman epistemology. *Social Epistemology, 26*(3–4), 463–469.

Mies, M., & Shiva, V. (2014). *Ecofeminism*. New York: Zed Books.

Plumwood, V. (1993). *Feminism and the mastery of nature*. New York: Routledge.

Plumwood, V. (2002). *Environmental culture: The ecological crisis of reason*. New York: Routledge.

Shils, E. (1981). *Tradition*. Chicago: The University of Chicago Press.

Snaza, N., Sonu, D., Truman, S. E., & Zaliwska, Z. (Eds.). (2016). *Introduction: Re-attuning to the materiality of education in pedagogical matters: New materialisms and curriculum studies* (p. xv–xxxiii). New York: Peter Lang.

Tuana, N. (2008). Viscous porosity: Witnessing Katrina. In S. Alaimo & S. Herman (Eds.), *Material feminisms* (pp. 188–213). Bloomington, IN: Indiana University Press.

Warren, K. J. (1997). Taking empirical data seriously: An ecofeminist philosophical perspective. In K. J. Warren (Ed.), *Ecofeminism: Women, culture and nature* (pp. 3–20). Bloomington, IN: Indiana University Press.

https://www.lllusa.org.

7 A Curriculum of the Commons
Learning in Detroit and Beyond

Rebecca A. Martusewicz

Twice before the publication of his book, *Revitalizing the Commons: Cultural and Educational Sites of Resistance and Affirmation* (2006), Chet Bowers visited Southeast Michigan, met with a diverse group of educator activists and with them toured a range of sites in Detroit where commons-based activities were (and are) alive and well. Forced subsistence in the wake of corporate abandonment pushed Detroit residents to develop and teach a complex system of self-determination and mutual aid. Their ongoing work addresses food insecurity, unemployment, the collapse of infrastructure, blighted housing, defunding of schools, and other effects of economic collapse. Community-based agriculture, exchanges of work for work, revitalization of parks, and common spaces, shared elder and child care, informal teaching and learning spaces in churches and neighborhood community centers, as well as the re-inhabitation of wildlife (pheasants, fox, deer, and other critters) in urban spaces create a renewed sense of what education could mean in these marginalized post-industrial neighborhoods (Bowers, 2016, p. 104). In this chapter, I offer a close look at what Chet hoped for as we explored these revitalization efforts and the commons-based relationships that grew even in as unlikely a place as Detroit, MI.

Defining the Commons

Drawing from the concept utilized in pre-Roman Britain, Bowers emphasized the commons as the life-sustaining practices of diverse cultures across the globe in relationship with their specific biospheres. He specified that the environmental commons included our non-monetized or commonly shared relationships with all aspects of the ecosystems within which we live: the air, water, forests, soil, living creatures, rocks, minerals, and so on. The cultural commons are defined by the practices, rituals, knowledge, traditions, and relationships (also non-monetized) by which people organize and experience daily life. Bowers was particularly interested in how these relationships and practices help to keep communities and the more-than-human world (Abram, 1996) healthy.

He argued that while we need to identify the existence and intergenerational importance of such practices in our modern culture, we should also look to diverse land-based or Indigenous cultures for the most significant examples from which to learn. The cultural commons include food ways (e.g., cultivation, harvest, and preparation), craft skills (building, weaving, or pottery), animal husbandry, medicinal plant knowledge, storytelling, music and other arts, games, healing arts, child and elder care, rules for decision-making, and so on.

Bowers believed strongly that "identifying the natural and cultural commons… provide(s) the conceptual basis for recognizing the aspects of the commons still viable in modern, technological and consumer-dependent culture" (2006, p. 2). He wanted teachers, teacher educators, and community educators to take seriously that these local and diverse forms of knowledge were extremely important in altering the current destructive epistemological basis of industrial market-based societies; not as a means of "going back" to some romantic notion of life once lived but rather, to expose valuable inter-generationally passed skills and knowledge that focus on caring for one another and the land. Students should learn basic skills and knowledge about how to develop relationships that encourage restraint and humility, mutual aid, kindness, neighborliness, and modes of meaningful work as essential values that keep communities sustainable and just. Emphasized here is that the commons is fundamentally about creating relationships whereby people care for one another and the natural world. The skills, practices and knowledge that Bowers emphasized could not be enacted outside of specific forms of relationality. We will see below how the learning we all did together in Detroit and beyond required the creation of lasting value-laden relationships and collaboration.

Bowers also emphasized the ways modern societies have historically undermined such values by enclosing the commons—making commons-based activities or relationships into commodities, accessible only with money. Enclosure or privatization through private property is a central aspect of the historical development of capitalism and industrialization, leading to the destruction of the planet in the name of inevitable human progress.

> In the modern industrialized West, enclosure has taken on new forms of expression that have radically reduced the cultural and environmental commons. This, along with the dominant belief system that privileges the interests of the individual over all other relationships, has become a major reason for the rapid degradation of the environment—and to the undermining of the self-sufficiency of other cultures.
>
> (Bowers, 2006, p. 3)

Understanding Bowers' commitment to exposing the deeply held assumptions and habitual practices that commodify nearly every aspect of our day to day lives helps make his interest in what was happening in Detroit—posterchild for blight, white flight and corporate abandonment—even more important (Bowers, 2016).

A Timely Meeting

I met Chet Bowers in 2000 at the annual meeting of the American Educational Studies Association in Vancouver, BC. I had been reading his work, in particular his book *A Culture of Denial* (1997), at the same time I was in conversation with a colleague from Detroit about intersecting social and ecological damages occurring in his own neighborhood and across the city more generally. Bowers' book along with his talk in a session called "Eco-Justice and Global Ethics" put me on the edge of my seat. He offered a poignant summary of how studying the primary languaging systems in our modern industrial culture, helps to expose the roots of intersecting ecological and cultural crises. Drawing on the work of anthropologist Clifford Geertz (1973) and others, Chet argued that culture should be understood as a system of signifying codes that produce patterns of belief and behavior articulated in everyday life as well as within institutional organization including official curriculum in schools. He made clear that this view of culture makes it possible to see important implications of modernist thinking and the associated meaning systems for both social and environmental systems. He was also adamant that this modernist worldview and its internalized and institutional effects are not the only cultural views or systems in the world. In fact, in *A Culture of Denial* (1997) he stated that "this more complex view of culture also makes it easier to bring into the discussion of ecological sustainability a comparative perspective that is less distorted by the modern bias against more traditionally oriented cultures" (pp. 25–26). While an early statement, this second dimension is an example of Bowers' commitment to examining diverse cultural commons, in particular, those within traditional cultures whose dominant patterns exemplify strong practices of care and connection.

As I listened to Chet that day, I began to wake up to something I had not really considered before; that there might be practices and relationships we engage daily in our own communities that could be identified and revalued to teach us how to get out of the mess our industrial system was making of the world. I thought about what was going on at home in Southeast Michigan, in particular in Detroit, as a hopeful example. Chet met me at the door after his session: "You want to talk?" Oh boy, did I want to talk! That was the start of a scholarly and pedagogical partnership that changed my life.

Faculty Retreats on EcoJustice and the Commons

Chet wanted a way to bring his ideas more directly to education faculty and community activists. He emphasized the mediating role of educators—professors, teachers, and community educators—in the reproduction of culture. If deeply held and linguistically passed on assumptions are not examined, and thus remain taken for granted, he argued, the crises we face will worsen. On the other hand, if teachers ask students to analyze their culture, in particular the dominant signifying patterns, wholly different ways of thinking and behaving could open up or be revealed. In short, it matters what is being mediated in these pedagogical processes. Chet wanted faculty to examine harmful root metaphors in their own teaching and create openings for exploring the commons. So immediately after the first meeting in Vancouver, we began to correspond, and with his student Jeff Edmundson, organize EcoJustice retreats. Faculty, graduate students and community activists from across the country were introduced to Chet's theoretical framework as it applied to our particular work. We shared what we were all trying to do to address growing social-cultural and environmental degradation with teachers and other educators in our communities.

The first retreat in Northern Michigan was held in the summer of 2002. It was organized by a series of morning lectures by Chet on the primary root metaphors and their importance in creating a mindset necessary to industrial capitalism. He was very methodical in defining each of several concepts (individualism, mechanism, progress, anthropocentrism, ethnocentrism and so on), offering poignant examples of how they intersect to form destructive taken for granted assumptions within all sorts of day to day interactions. In the afternoon, other related topics were introduced by participants including ecofeminism and the intersection between gender and ecological domination, and a critique of urban impoverishment via racism and corporate irresponsibility. This latter discussion was introduced by my colleague, Detroit activist and Eastern Michigan University (EMU) professor, Charles Simmons.

I had been working with Charles and a group of Detroit community activists to clean up an abandoned factory lot in his neighborhood. At this early retreat, Charles shared the historically racist, economic, and social policies that had devastated the city of Detroit. In 1967 a riot broke out (called a rebellion by African American residents) that was a reaction to specific problems suffered by the Black community since their arrival to Detroit via the Great Migration: police brutality, segregated, and inferior housing, unequal treatment on the job, rising unemployment as work was outsourced to the suburbs, or the south, and eventually out of the country. Charles talked with us about the on-the-ground activism among mostly Black residents of Detroit working together in neighborhoods to care for one another. A particularly skilled and engaging mediating educator, Charles described the patterns of intergenerational care, traditional values of self-determination, love, kindness, generosity of spirit, interdependency, and mutual aid—the building blocks of the commons. The discussion he encouraged at that first faculty retreat brought to the fore both sides of Chet's analysis: the rooted assumptions driving all sorts of impoverishment in the name of progress, and the only way out of it: through organized acts of mutual aid, shared practical skills, and kindness.

The Detroit activists that I met that same summer, led by Charles, met every Saturday morning to clean up a mess created when a construction company illegally dumped piles of concrete, dirt and other debris on a former factory site right across the street from Charles's house. The group included neighbors as well as members of a group calling themselves the "Committee for the Political Resurrection of Detroit—CPR Detroit." CPR was made up of volunteers committed to making this and other neighborhoods safer and more functional for their children both via projects like this one but also by working on policy in the city. As they pushed wheelbarrows filled with dirt, rocks, and chunks of concrete, they shared hopes to build a basketball court and playground here, as well as raised-bed gardens. They described what the neighbors were doing for and with each other to address day to day needs: exchanges of work for work like sharing a running car for a fixed roof, daily check-ins and help bringing groceries to housebound elders, cultivation of backyard gardens and sharing of food. An elder woman named Sammy shared old wisdom passed down from her southern ancestors about the medicinal properties of plants growing in the cracks of sidewalks; the ways she used her tiny backyard to grow all sorts of flowers (a beautiful array!) as well as a wide variety of vegetables, which she shared freely with all who needed to eat. There was laughter and storytelling as people worked together on those Saturday mornings and I felt immense joy and warmth from them. Though I did not have this language at the time, I was experiencing what WinklerPrins and DeSouza (2005) call "an economy of affection."

Putting their stories in the context of Chet's emphasis on diverse and ancient commons-based knowledge, I saw the wisdom imbedded in their practices and relationships as a particular kind of curriculum shared with me and others taking part in the work. Before connecting Chet's work to these Saturday mornings, I had never really imagined what subsistence and conviviality among cooperative neighbors and interested city folk meant, what its value was relative to more official forms of knowledge learned in schools and universities. At this point, much of what I read from Chet's work was focused on what we could learn from Indigenous people and their diverse cultural ways. But this was Detroit, once a beautiful and highly successful example of modern industrial success and prowess, now a canary in the coal mine of post-industrial decline and impoverishment (see Sugrue, 2005). It suddenly became clear to me. The people Charles introduced me to, and others I was soon to meet, were challenging material and ideological impoverishment head on by developing and teaching each other what they needed to live well together. This work was about love. Their explicit care work demonstrated clearly the dangers of individualism and the myth of constant profit-oriented growth as necessary progress that Chet examined. Here were explicit local examples of people engaging the commons—ancient knowledge, skills and relationships brought to a very modern context—as the primary solution to the damages caused by industrial irresponsibility. I was keen to share with my students and with Chet what seemed to me to be a modern model that we could all learn from.

A Symposium on Revitalizing the Commons of Southeast Michigan

In 2003, after two successful faculty retreats (the first in Michigan, the second in Miami, FL), we began thinking about organizing an event with Detroit activists. Chet's thinking and writing became even more focused on the commons as a way to make more concrete his interest in community interdependencies. A key influence in both his written work and in the teaching we experienced with him was Gregory Bateson's *Ecology of Mind* (1972). Bateson's analysis reveals the ways complex patterns of communication (what he called cybernetics) worked as circuits made of differences that make a difference in both cultural and natural systems. This insight, along with Bateson's recognition of the metaphorical nature of linguistic systems, were central to Bowers' cultural critique as well as his recognition of cultural patterns that offer sustainable ways of being (Bowers 1993, 2011). With these concepts as tools, I saw vital differences taking shape in the relationships among the Detroit activists and the neighborhoods where they worked. Essential limits were being respected and practices of care mattered, sometimes in life or death ways (see Martusewicz, 2009, 2016). Seeing the connections so vitally in my own local context, I wanted to share the Detroit example with Chet and others directly. And I wanted to offer Bowers' insights to those I was working with.

In the spring of 2004, Johnny Lupinacci (then enrolled in a Master's Degree in Social Foundations of Education at EMU) and I organized a one-day symposium called Revitalizing the Commons, inviting a whole range of the most prominent Detroit activists including among others, Jim Embry, Elena Herrada, Charles Simmons, and Grace Lee Boggs, as well as others from the Ypsilanti/Ann Arbor area. Teachers, community activists, teacher educators and students gathered around a table at EMU for a seminar designed to introduce our work to each other in the hopes of energizing some common projects. Chet opened the discussion, grounding his introductory remarks in a description of what respecting traditional knowledge in the African American community might look like especially in a place like Detroit. We then invited the folks around that table to describe for us their work in revitalizing those practices and the effects of their work. It was a heady day for all of us. The conversation that resulted was mostly introductory and somewhat tentative, focusing mostly on who each participant was, what their work was focused on, and what we might do together, should we find a way to move forward. We were all aware that this could either be a beginning or a waste of valuable time.

While there were limits to a one-day symposium, the day opened the possibility of building relationships among activists and educators, and among Detroit folk and my students (mostly white and from the suburbs). And it served as an important way for Chet and me to bring this emphasis on the commons down to earth in a specific urban context. Embarking on this particular work, our friendship deepened, and Chet's ideas began to take concrete form in the conversations, actions, and relationships that my students and I made with these Detroiters.

Exploring the "Patterns that Connect" in Detroit

With these relationships established and growing, we followed up the Revitalizing the Commons Symposium with a conference held the next fall in conjunction with an event organized in Detroit by local social welfare activists. As a precursor to the conference, we arranged for a tour of some of the sites in the city where commons-based work (particularly in urban agriculture and the arts) was most visible. This was by far the most important piece of the weekend event for all of us, but perhaps most specifically for Chet.

For Chet, the "patterns that connect" and the essential "relationality" in the network-building and friendships growing among my students, these Detroit activists/educators and him was particularly important. As mentioned earlier, patterns of connection and relationality form the bonds that hold us "in common," and, if nurtured, thus mutually responsible for one another. As Wendell Berry (2010) argues, they form an *order* that indicates important limits and requires respect and responsibility. The making of these bonds is active, happening in all manners of exchange and shared labor: working, talking, laughing, grieving, and loving one another.

Capitalism and other modern forms of industrialization rely on the illusion that what matters most is individual success and power, that the most viable civilizations are hierarchized around the individual accumulation of more and more resources that allow for the domination of all other aspects of life as those resources are gathered by some and withheld from others. Such accumulative "progress" is the source of never-ending technological development that further mechanize once living relationships and processes in order to make the production of wealth and power more and more "efficient" for those who benefit. Impoverishment of those who are the losers in this selfish system is thus normalized as a natural outcome of inherent strengths and weaknesses. Enacted to structure and enclose day to day life, these assumptions are carried through discourses that are institutionalized, breaking the bonds necessary for communities to care for one another and flourish. Such damage is how we come to find ourselves now facing the most serious existential threat as the climate warms and whole continents burn.

Bonds of love and devotion, now too often erased or inferiorized as weakness are actually the source of true happiness and life itself (Akulukjuk, Erkaeva, Rasmussen, & Martusewicz, 2020; Martusewicz, 2019). In these retreats, symposia, conferences, classrooms, and informal meetings, we built friendships and deep connective relationships, both locally and beyond. These relationships of affection and commitment, kindness and joy, built within serious pedagogical, theoretical and activist work formed the basis of an educational movement. Most of these friendships have endured, creating an even stronger network of scholars, activists, and educators, some of whom are included in this text as editors and authors.

What is most important is that our work together is fully dependent on these nurtured relationships. This process was not without serious mistakes along the way, where good strong friendships were harmed or even broken. But what I want to argue now is that no matter what aspect of life one is considering—that which brings about other species of animals and plants, whole biomes, or human life—there is no creative activity without the proliferation of generative relationships among the various elements or actors at play (Martusewicz, 2016). The quality of those relationships and the differences made through them (large or small and mostly unseen) matters. They are always fragile. How they are tended either allows life to flourish or can cause serious impoverishment of both spirit and material well-being, and ultimately if left without care, system collapse.

This is the fundamental lesson offered to us by scholars like Gregory Bateson (1972), Wendell Berry (1996), Chet Bowers (2006), Val Plumwood (2002), Susan Griffin (1995), and many others who have been primary teacher-leaders in this movement. What I experienced with such joy in Detroit was a result of people who understood this in the most deeply practical ways. They came together to enact the commons in the face of devastating poverty, pollution, racism, food insecurity, and continued economic exploitation. It was also an important aspect of the friendship that I had with Chet. He understood this essential vulnerability. And, although he was often misunderstood due to his propensity for abstract academic language in his writing and also his tense discomfort within some settings, he spent his life dedicated to making others understand this fundamental concept.

The Day in Detroit

When we went into the city that fall day in 2004 with Jim Embry as our guide, Chet experienced firsthand what I had been trying to describe to him. Detroit is a complicated place. There is a wide range of discourses at play including Marxism and deeply held, historically grounded Black civil rights arguments, activists who rely on Freirean theory and critical pedagogy, agricultural science blended with Afro-centric philosophy, union politics, radical feminism, a variety of religious perspectives, even semi new age spiritualism—all mixed together to frame the work that I had been witnessing. As the day unfolded, Johnny and I watched with delight as Chet waded right in, engaged in all sorts of conversation, listened intently and learned from teachers, artists, neighborhood gardeners, even Capuchin monks, with grace and sheer happiness. It was as if he had known these folks forever. He was relaxed and open, totally in his element taking it all in with a kind of gusto that I had not really seen until that day. My respect for him as my friend and mentor deepened with it.

We visited the site in the neighborhood where my journey had begun, talking with Charles who also shared with us a museum of African American history, politics, and activism set up in his house (later called the Hush House). We talked with neighbors, Johnny and Sammy, who were working on a community garden just down the street. We walked on a block-long cement slab, all that was left of the munitions factory building where we had worked to clean up illegally dumped debris a few years prior. Charles and his neighbors talked about a city-wide project to increase urban agriculture being supported by an organization called *The Greening of Detroit,* which had set up a network of composting sites around the city where gardeners could access good rich soil and offered help testing the soil where they planned to garden.

We also visited *EarthWorks* where Capuchin monk Rick Samyn had started a very productive vegetable farm on less than an acre of land, providing a soup kitchen run by the monks with literally tons of fresh vegetables every season. *EarthWorks* is still the site of strong hands-on urban agriculture education for both adults and children in after-school programs. The organization now works in partnership with the James and Grace Lee Boggs School to introduce young children to the principles and values needed for creating a sustainable Detroit (see https://www.cskdetroit.org/earthworks).

We saw a model for this sort of education in action that day in 2004 when we went to Catherine Ferguson Academy (CFA), a public school dedicated to pregnant and parenting girls. The biology teacher Paul

Weertz, with the support of principal and founder Asenath Andrews, had started a farm on the site of the school with his students. It had begun as an alternative to exposing the girls to formaldehyde used in frog dissection, which would likely harm their unborn babies. And now there was an orchard, vegetable gardens, chickens, goats, a horse, rabbits, even a steer at one point. The girls had raised a barn, helped Paul cut and bale hay in the summer, and were the primary caretakers of the animals and the cultivation of the gardens. They were also renovating houses in the neighborhood. They showed us around the grounds and school which also housed a nursery and pre-school for the girls' children. We had time to hear Paul and Ms. Andrews talk about the goals and successes of this unique school. Unfortunately, CFA closed in 2014, having been sold to for profit charter organization as part of Detroit's school defunding program, and ultimately losing its primary goal of helping pregnant and parenting girls, and with it, enough enrollment to remain viable.

There have been many changes in Detroit since 2004, as the bankruptcy of the city gained national attention. The state appointed anti-democratic Emergency Managers to gut the Detroit Public School System and the pensions of city workers, and sell off public resources like parks, schools and other resources that had been paid for by the Detroit taxpayers. Water shut-offs, illegal foreclosures and a refusal to fund basic infrastructure needs (lights, water and sewer, roads, schools) in neighborhoods deemed to be unwise "real estate investments" have paved the way for "urban renewal" processes that are pushing more and more people (primarily African American families who stayed after white flight and economic abandonment) out of their homes and the city (Kurashige, 2017; Pedroni, 2011).

But that day and through the weekend, we explored the roots of real change for and by the people of Detroit. We uncovered their particular curriculum of the commons—the ideas and principles, practices, and relationships that they were teaching one another and their children in order to address day-to-day needs in ways determined by themselves, not the state or the city government. The conference hosted some of the same folks that we had visited the day before who came to share their stories and advocate for the specific sorts of revitalization efforts needed to support sustainable life in the city. Before we finished our conversations, Chet promised to fund a tool-sharing "library" where people could access shovels, rakes, wheelbarrows, hoes, and hand tools essential for cultivating much needed healthy food. I deepened my friendship with Jim Embry, along with Chazz Miller and Aurora Harris, a muralist and poet who together established the Artist Village on the northwest side of the city. We later created a free summer camp on art, poetry, and gardening for Detroit children funded by a Department of Justice grant through EMU.

Soon after the 2004 weekend, Chet published his book, *Revitalizing the Commons: Cultural and Educational Sites of Resistance and Affirmation* (2006), which included a chapter on Detroit (Bowers & Martusewicz, 2006). We wrote together on the commons again for Provenzo's *Handbook of the Social Foundations* (Bowers & Martusewicz, 2009) and we created the EcoJustice Dictionary with a group of graduate students, originally published on a website called EcoJustice Education, created by Chet, Johnny Lupinacci, and me. The site also housed an open source journal called *The EcoJustice Review* (no longer available). The EcoJustice Dictionary is now published on Chet's website at https://cabowers.net/CAdictmain. php. Finally, we organized one last retreat in Northern Michigan during the summer of 2006, to expand, yet again, the web of relationships needed for this work. That retreat focused exclusively on the commons. Chet did some lecturing to introduce central concepts, but there was more exploring of what examples of the commons could be found in our daily lives, and how we could engage that knowledge with students.

The Years and Work Since

What's New in Detroit?

Detroit's currently heralded "renewal" dubbed New Detroit is quite literally a program of dispossession by accumulation as people have lost their neighborhood schools, jobs, and homes to downsizing and gentrification strategies. Scott Kurashige (2017) calls this program a neoliberal counter-revolution, one to combat the "Fifty-Year Rebellion," "a strong web of interconnected grassroots organizations, civil rights, and environmental activists who have been working since the 1960's to demand more equitable treatment of residents (85% African American) and to revitalize their neighborhood commons" (Martusewicz, 2005, 2019; see also Boggs, 2011; Lupinacci, 2017). This quiet rebellion is what we witnessed and engaged that day together in Detroit. Today, in spite of the concerted efforts to quell its successes, it is stronger and even more organized. It represents a renewal quite different than the one being given official attention.

Of particular note is the Detroit Black Community Food Security Network (www.DBCFSN.org), led by Malik Yakini. Yakini was a founder and principal of Nsoroma Institute, one of two Afro-centered schools in Detroit, now closed due to the loss of its charter. DBCFSN began as people from across the city came together to begin discussing the intersections among racism and food insecurity. Over two years a diverse group of activists, educators, and farmers crafted a food policy for the city and established a 7-acre farm on the northwest side of the city called D-Town Farm. Vegetable plots, greenhouses, orchards, mushroom cultivation, and an apiary comprise this now internationally known urban farm. The food policy, eventually adopted by the Detroit City Council, included strong language about the need for youth education and D-Town has become the site for the cultivation of more than just food. Young men and women known as Food Warriors come to the farm to learn ancient African principles associated with creating strong sustainable, loving communities that start with growing healthy food.

With D-Town as one of the strongest examples, Detroit's urban agriculture movement has grown one hundred fold since the years Chet visited. A network of grassroots organizations supports both backyard gardens and other urban farms providing vegetables and greens for local markets and restaurants. I continued to take my students into Detroit to learn from these folks and to invite them to teach us at EMU through the end of my position there. Detroit urban agriculture continues to offer the strongest curriculum of the commons alive and active that I know of.

My Work as a Tribute to Chet Bowers

With this model from Detroit and the people involved in developing it as primary mentors, I must say that the most fruitful result of this work with Chet Bowers has been the confidence we gained to teach from and develop the framework and theory that he laid out over the course of his academic life (see Martusewicz & Edmundson, 2005). Using Bowers' fundamental ideas and what we learned together in Detroit, Johnny Lupinacci and I worked very closely to develop EcoJustice curriculum for both undergraduate and Master's level teacher education at EMU. Most of our students were from the working-class suburbs of Detroit, so our experiences with the people in the city became very germane to our pedagogy. We were enacting what Chet had hoped for in the faculty development retreats and collaborative work with us over the years. And eventually, Jeff Edmundson, Johnny and I used these experiences to write a textbook laying out the primary ideas in that curriculum. *EcoJustice Education: Toward Diverse, Democratic and Sustainable Communities* (Martusewicz, Edmundson, & Lupinacci, 2011, 2021) allowed us to articulate Bowers' ideas along with our own translation and interpretation of those ideas within a frame that made sense in the teacher education programs where we taught (and where Johnny still teaches). Increasingly we developed clearer understanding of the essential framework that Chet had initially offered: (1) A cultural ecological analysis of the deeply rooted discourses comprised of what Bowers called root metaphors creating beliefs and behaviors, institutions, systems, and day to day interactions that are harmful to both social and ecological well-being; (2)the identification and revitalization of the world's diverse cultural and ecological commons in order to find sustainable ways of living; and (3) the development of the imagination necessary for both of these practices via pedagogies of responsibility.

Now in its third edition (2021), that text has opened many more doors for the three of us than would have been possible without it or Chet's initial guidance. Chet's influence helped me to see the importance of taking this movement more directly into schools and to practicing teachers. I worked for two years, 2006–2008, with teachers at Souhegan High School in Amherst, NH where we were able to develop a senior seminar in food systems and sustainability. That work resulted directly from my experiences with urban agriculture in Detroit with Chet and the others. I used that experience with two other strong women educators to create the Southeast Michigan Stewardship Coalition to help form partnerships between schools and community organizations in order to support teachers and their students to address real community problems and begin to revalue their local commons (Lowenstein, Voelker, & Martusewicz, 2010).

I continued to explore both the Detroit context as well as the commons of my childhood and lessons learned from my family, especially my mother (Akulukjuk et al., 2020; Martusewicz, 2014). As I turned my attention to my hometown in rural Northern New York, I became more and more interested in thinking about the particular problems manifested historically in rural areas. As I had in Detroit, I was looking to ground the analysis Chet had offered us in the everyday lives of people in the places where they live.

In 2015, I met Wendell Berry, whose work Chet had introduced to me many years prior. Working on a book about Berry's influence, he invited me to spend an afternoon talking together at his farm and then corresponded over several years about the ideas I was trying to articulate in that book (Martusewicz, 2019). Those conversations along with Berry's incredible oeuvre have been life changing for me. Now in my retirement, feeling a bit of freedom to shift the form my scholarship takes, I wish to bring his insights to a family memoir of my grandfather's life as a dairy farmer in Northern New York where I grew up. As I clear my desk of other commitments, this ode to Chet Bowers included, I will take my attention there to the history of agriculture in New York State and the rural commons that my grandparents engaged in their earliest years together. While this turn is certainly different from exploring the neighborhoods of Detroit, I am extremely grateful for all I learned there among such amazing teachers and activists. And thinking back to that time with Chet has me once again recognizing with deep gratitude all that he taught me. Our years of close association and friendship were themselves a multifaceted curriculum of the commons.

References

Abram, D. (1996). *The spell of the sensuous: Perception and language in a more-than-human world.* New York: Pantheon Books.

Akulukjuk, T., Erkaeva, N., Rasmussen, D., & Martusewicz, R. A. (2020). Stories of love and loss: Recommitting to each other and the land. In H. Bai, D. Chang, & C. Scott (Eds.), *Ecological virtues*. Regina: University of Regina Press.

Bateson, G. (1972). *Steps toward an ecology of mind*. Chicago: University of Chicago Press.

Berry, W. (2010), *What matters: Economics for a renewed commonwealth*. Forward by H. E. Daly. Berkeley: Counterpoint Press.

Boggs, G. L. (2011). *The next American revolution: Sustainable activism for the twenty-first century*. S. Kurashige (Ed.). Berkeley: University of California Press.

Bowers, C. A. (1993). *Education, cultural myths, and the ecological crisis: Toward deep changes*. Albany: State University of New York Press.

Bowers, C. A. (1997). *The culture of denial: Why the environmental movement needs a strategy for reforming universities and public schools*. Albany: State University of New York Publishers.

Bowers, C. A. (2006). *Revitalizing the commons: Cultural and educational sites of resistance and affirmation* (pp. 47–84). Lanham: Rowman and Littlefield.

Bowers, C. A. (2016). *Reforming higher education in an era of ecological crisis and growing digital insecurity*. Claremont: Process Century Press.

Bowers, C. A., & Martusewicz, R. A. (2009). Eco-justice pedagogy and the revitalization of the commons. In E. Provenzo, Jr. (Ed.), *Encyclopedia of the social and cultural foundations of education*. Los Angeles: Sage Publishers.

Bowers, C. A., & Martusewicz, R. A. (2006). Revitalizing the commons of the African-American communities in Detroit. In C. A. Bowers (Ed.), *Revitalizing the commons: Cultural and educational sites of resistance and affirmation* (pp. 47–84). Rowman and Littlefield, Lanham: MD.

Bowers, C. A. (2011). *Perspectives on the ideas of Gregory Bateson, ecological intelligence, and educational reforms*. Eugene: Eco-Justice Press.

Edmundson, J., & Martusewicz, R. A. (2013). "Putting our lives in order": Wendell Berry, ecoJustice, and a pedagogy of responsibility." In A. Kulnieks, K. Young, & D. Longboat (Eds.), *Contemporary studies in environmental and indigenous pedagogies: A curricula of stories and place*. Rotterdam: Sense Publishers.

Geertz, C. (1973). *The interpretation of cultures: Selected essays*. New York: Basic Books.

Griffin, S. (1995). *The eros of everyday life: Essays on ecology, gender and society*. New York: Doubleday.

Kurashige, S. (2017). *The fifty-year rebellion: How the U.S. political crisis began in Detroit*. Oakland: University of California Press.

Lowenstein, E., Martusewicz, R. A., & Voelker, L. (2010). Developing teachers' capacity for ecoJustice education and community-based learning. *Teacher Education Quarterly*, Fall, 99–118.

Lupinacci, J. (2017). Resistance, wisdom, and grassroots education: Lessons from Detroit. In W. T. Pink & G. Noblit, (Eds.), *Second International handbook of urban education* (pp. 833–851). Cham: Springer International Publishing.

Martusewicz, R. A. (2005). Eros in the commons: Educating for eco-ethical consciousness in a poetics of place. *Ethics, Place & Environment, 8*(3), 331–348. doi: 10.1080/13668790500348299.

Martusewicz, R. A. (2009). Toward a "collaborative intelligence": Educating for the cultural and ecological commons in Detroit. In M. McKenzie, P. Hart, H. Bai et al. (Eds.), *Fields of green: Re-storying education* (pp. 251–270). New York: Hampton Press.

Martusewicz, R. (2013). Toward an anti-centric ecological culture: Bringing a critical ecofeminist analysis to ecojustice education. In A. Kulnieks, K. Young, & D. Longboat (Eds.), *Contemporary studies in environmental and indigenous pedagogies: A curricula of stories and place* (pp. 259–272). Rotterdam: Sense Publishers.

Martusewicz, R. (2014). Letting our hearts break: On facing the hidden wound of human supremacy. *Canadian Journal of Environmental Education, 19*, 31–46.

Martusewicz, R. A. (2016). Reading Bateson and Deleuze on difference: Toward education for eco-ethical consciousness. In W. M. Reynolds & J. A. Webber (Eds.), *Expanding curriculum theory: Dis/positions and lines of flight* (2nd ed.). New York: Routledge.

Martusewicz, R. A. (2019). *A pedagogy of responsibility: Wendell Berry for ecojustice education.* New York: Routledge.

Martusewicz, R. A., & Edmundson, J. (2005). Social foundations as pedagogies of responsibility and eco-ethical commitment. In D. Butin (Ed.), *Teaching context: A primer for the social foundations of education classroom* (pp. 71–92). Mahwah: Lawrence Elrbaum Publishers, Inc.

Martusewicz, R. A., Edmundson, J. & Lupinacci, J. (2011). *EcoJustice education: Toward diverse, democratic, and sustainable communities.* New York: Routledge.

Martusewicz, R. A., Edmundson, J. & Lupinacci, J. (2021). *EcoJustice education: Toward diverse, democratic, and sustainable communities* (3rd ed.). New York: Routledge.

Pedroni, T. (2011). Urban shrinkage as a performance of whiteness: Neoliberal urban restructuring, education, and racial containment in the post-industrial, global niche city. *Discourse: Studies in the Cultural Politics of Education, 32*(2), 203–215.

Plumwood, V. (2002). *Environmental culture: The ecological crisis of reason.* New York: Routledge.

Sugrue, T. (2005). *The origins of the urban crisis: Race and inequality in post-war Detroit.* Princeton: Princeton University Press.

Winkler-Prins, A. M. G. A., & DeSouza, P. (2005). Surviving the city: Home gardens and the economy of affection in Brazilian amazon. *Journal of Latin American Geography, 4*(1), 107–126.

https://www.cskdetroit.org/earthworks.

https://cabowers.net/CAdictmain.php.

8 Lessons from a Pandemic

Can We Reclaim Our Cultural and Environmental Commons?

Susan Huddleston Edgerton

In a series of pandemics brought about by European invasion of the Americas, indigenous populations were reduced by 90 percent. Based on an estimated indigenous population of 60 million, that constituted a 10 percent reduction of the global population, the equivalent of 780 million people in today's terms. That loss forever changed life in the Americas (Koch, Bierley, Maslin, & Lewis, 2019). Small ill-prepared bands of Europeans would never have overwhelmed some 60 million indigenous on this continent without the help of—mostly unintentional—biological warfare. This forever changed life on the planet. We will never know precisely how it was changed but can speculate that the ancient knowledge of the environment that died with those indigenous might have been incredibly useful to the rest of us.[1]

Now, on a planet of 7.8 billion, the COVID-19 pandemic is global. No one came to it with immunities because the virus is novel, having jumped from animal to human in a way that points toward our increasingly troubled relationship with animals. We now live in a world where any new virus that takes hold in a population is likely to move quickly around the planet because of the ways we travel and trade. As long noted by Bowers (2001) and others, the way we travel, eat, work, consume, and entertain ourselves have also significantly contributed to the warming of the planet. In that way, at the least, the pandemic and the climate crisis are connected. The emergence of this pandemic and its effects so far have clearly demonstrated the dangers of not recognizing the interconnectivity and relational patterns of being in the world. In effect, we have forgotten just how the world works (Bowers, 2018). And though it seems unlikely at this point that the pandemic will kill 10 percent of the global population—at this writing, 809,000 are dead worldwide—it will, as in the world during European colonization of the Americas, forever changed our lives.

In this chapter, I will explore what the pandemic and its collision with this political era is teaching us about ourselves and our environment—how we must change and how our awareness of capacity for change has evolved through the related crises of pandemic, climate change, and political disarray. So much of Chet Bowers' work speaks directly to this inquiry, so I contemplate the ways he might have responded to the current crises were he still among us. Finally, I argue, as I believe Bowers would, that our educational institutions have a unique opportunity to reimagine curriculum and instruction in the interest of preserving and revitalizing our environmental and cultural commons.

Much of what Bowers wrote in 2016—the last years of his published works—was not prophecy but description and analysis of what was already happening. Although he mentioned the likelihood that new germs to which we have no immunity will appear, global pandemic was not in the center of his radar. Bowers died in 2017, just under six months after the inauguration of Donald Trump, and so was unable to observe the collision of ineptitude, self-serving pathology, historical ignorance, and corruption in the White House with the worst global pandemic in one hundred years. These two phenomena—Trump and pandemic— are shining bright light on many of the very issues and concerns about which Chet Bowers spent decades writing. In this he was prophetic.[2]

Foundational Conceptual Themes

Conceptual themes from the body of Bowers' work to be developed in context for this writing include: (1) The need to revitalize the *cultural commons* and through that, the *environmental commons*; (2) the ways in which language, especially via *root metaphors* we inherit, reproduces ways of thinking and knowing that came into being before we understood that we face ecological limits; (3) the impact of *expanding reliance on digital technologies* on both cultural and environmental commons; (4) the complicated and contradictory *role of experts* in our efforts to reclaim and revitalize the cultural commons; and (5) the role of consumer-dependent monetized economy in creating *deep inequities*, including abuse of the "more than human" world of plants and animals, that hinder our ability to reclaim the commons. I begin with some definitions and brief explanations before exploring these themes in the context of the pandemic.

Bowers explored the ways in which root metaphors we inherit reproduce ways of thinking and knowing that came into being before we understood that we face ecological limits. Root metaphors are "[t]he languaging processes [that] carry forward past ways of thinking that are based on assumptions unique to the culture" (Bowers, 2017). "Progress" is one example of a root metaphor. He writes, "[T]he idea of progress cannot be reconciled with exploitation of the environment" (Bowers, 2016, p. 6). We will see in the pages that follow that these sorts of root metaphors also lead us into the trap of double bind thinking—contradictions between explicitly held ideas and values. Of course, there are also root metaphors in linguistic systems of cultures that have learned to live sustainably. It is these traditions that we need to understand and embrace.

Resistance to linguistically sedimented ways of thinking and understanding that prevent us from recognizing ecological limits, Bowers has insisted, depends on reclaiming and revitalizing cultural commons that elevate relationships and traditions not dependent on consumerism and monetized economies. A revitalized cultural commons offers our best hope for also revitalizing the environmental commons, he suggests. Cultural commons, as Bowers understands, are the cultural patterns and traditions where intergenerational knowledge of how to best care for a community are honored and maintained; environmental commons are all those natural systems, water, air, soil, etc., that are shared without cost by communities (2017).

The digital explosion is a growing force in individualistic consumer-dependent culture, Bowers argues, leading to the loss of important traditions such as democracy, privacy, long-term memory, and meaningful work–that is, work that connects us to one another and the non-human environment (Bowers, 2010). Our increasing reliance on digital technologies will ultimately lead to greater violence amongst ourselves and toward our ecosystems, according to Bowers (Bowers, 2016). We are seeing that violence in the form of surveillance that invites public scrutiny and viral shaming, organization of violent domestic elements such as white nationalists, surveillance that enables unprecedented state violence against groups and individuals (Bowers, 2016b). This is not to suggest that we should eliminate digital technologies (even if we could). There are and will be ways these technologies enhance our lives and our prospects for survival. In this era of political chaos and pandemic, we are seeing the double face of digital technology and the ways it deploys root metaphors from our inherited linguistic milieu. One face accelerates root metaphors such as "mechanism" and "efficiency" that disguise harms to our environment and psyches; the other face enables new kinds of community and organizing, especially, for example, in the face of a pandemic where we must stay physically apart.

In sum, Bowers explored the consequences of the loss of connection to our past, to traditional ways of knowing that can sever our tethers to consumer-dependent monetized economies that are on a collision course with mass extinction, climate change, environmental degradation. He has delineated many of the ways that our new reliance on digital technologies drives us further away from our humanity, from our ability to do meaningful work that is ecologically based rather than destructive of our shared home, (Bowers, 2011b, 2011c, 2016a). His message has been clear and sustained. Still, the message has been slow to register. Few, it seems, are sympathetic to a Cassandra. But now we are living in the midst of a pandemic that is wreaking havoc on our economic system and our lives. It seems to have our attention. What will we do with that?

Pandemic Lessons

Food and Relationship to Animals

This coronavirus, evidence suggests, probably originated in a bat, then possibly moved to a pangolin, and from there into a human. Bats and pangolins are just a couple of the wild species that have been on display at a market in Wuhan, China, since prohibited, where the virus might have originated. Pangolins are not native to China but are typically poached from their native habitats, where they are increasingly endangered and exported to China, where they come into contact with animals they normally would not encounter, in this case certain species of bats. Bats carry coronaviruses but do not themselves become ill from them. The pressure placed on animals when we ravage their habitats and force them into places they do not belong, and into increasingly smaller places where human contact becomes more likely, is responsible for many of the zoonotic diseases that are passed to us. We also pass diseases to animals (Samuel, May 12, 2020). "Globally, an estimated 75 percent of newly appearing infectious diseases are "zoonotic" like this, meaning they can pass from non-human animals to people. Infectious-disease experts warn that nature harbors more than a million undiscovered viruses" (Peeples, 2020). The global trade in wildlife has been linked to the spread of other pandemics, including Ebola, SARS, and MERS. Contrary to some claims that the wildlife trade is rooted in traditional culture, the Humane Society reports that in China, for example, it was non-existent prior to the 1980s. Indeed, if there were a global movement to revitalize the cultural commons that enables us to live within our ecological means, as Bowers advocates, these practices would be unknown.

It must be understood, however, that the problem is not only located in the global wildlife trade. "The meat we eat [in the United States and wherever factory farming is practiced] is a pandemic risk, too" writes Sigal Samuel (June 10, 2020), and this is another reason to refrain from casting blame outside our boundaries. We can and have experienced homegrown global pandemics due to the practice of factory farming in which animals are crowded into small unsanitary spaces where they live tortuous lives. These farms provide around 99 percent of America's meat and similar facilities supply more than 90 percent of meat globally. There are factory farms in this country for cows, pigs, chickens, and fish. The animals have been bred to be genetically very similar for traits that enhance meat desirability, which facilitates the spread of viruses among the animals where the viruses can grow more virulent. Indeed, H1N1, or "swine flu" originated in North American pig farms, then became a global pandemic that killed hundreds of thousands of people. It is only a matter of time before other viruses or antibiotic resistant bacteria arise and hop to people from these practices.

The pandemic has highlighted our understanding of the importance of our diets to our health, the lives of non-human animals, and to the lives of humans tasked with processing meat from factory farms and working the fields of our fruits and vegetables many of whom are black and brown people. There were multiple outbreaks of COVID-19 in slaugherhouses due to the conditions under which people work. Farmworkers in the fields, many of whom are undocumented immigrants or guest workers from Mexico, are also at greater risk of infection. They live in cramped quarters without possibility of social distancing and have no health insurance or paid sick leave so must work even when infected. COVID-19 threatens our food supply chain at every level. Hence, we need to rethink the turn we took toward efficiency in agriculture and return to resilience. Resilience means doing more of what some communities are already doing—buying from local diversified farmers at local farmers markets, buying memberships in community-supported agriculture (CSA), reducing if not eliminating meat consumption, and buying any meat one eats from small local farmers (Pollan, 2020). Indeed, by some measures animal agriculture as practiced today is responsible for as much as half of all greenhouse gas emissions, depending on how it is measured (Foer, 2019).[3]

Denial and Mistrust in Experts

When the novel coronavirus (COVID-19) began spreading from China to other countries, officials at the Centers for Disease Control (CDC), epidemiologists and virologists, began warning US officials and citizens to prepare—that it was not a matter of *if* the virus would spread here but of *when*. The World Health Organization (WHO) warned in January 2020 that the "global risk was high." President Trump announced, "We have it totally under control. It's one person coming in from China, and we have it under control. It's going to be just fine" (Stevens & Tan, 2020). On February 10 at a rally in New Hampshire the president said, "Looks like by April, you know, in theory, when it gets a little warmer, it miraculously goes away" (Stevens & Tan, 2020). Though his commentary evolved to accommodate the situation as the virus rapidly spread, he rarely and barely acknowledged the advice of experts working within his administration, and even at one point accused democrats and "fake news media" of inflaming the situation for political purposes. In our deeply politically divided country—divisions greatly exacerbated by the president's rhetoric from the beginning of his campaign—his followers largely chose to believe him over health care experts and to downplay the significance of the pandemic. Even when it became clear that so many were falling seriously ill and many of them dying, the recommendation for social distancing and the wearing of masks in public places became politicized. After two months of orders to shelter in place and economic lock-down, there were protests across the nation at state capitals to "open up the economy, end the lock-down" by people carrying signs saying such things as "I need a haircut," and some carrying assault-style weapons. Conspiracy theories were rife on social media suggesting, for example, that China engineered the virus. Right-wing news outlets that are supportive of the president helped at times to amplify conspiracy theories and to downplay the seriousness of the pandemic.

Bowers rightly criticizes the language of academia, including scientists, for lack of awareness of the root metaphors that create double bind thinking, preventing us from developing ecological intelligence and from revitalizing our cultural commons that will enable us to move away from consumer-dependent monetized economies. The conceptual double bind is a concept that Bowers borrowed from Gregory Bateson, who explained it as "different from the nature of a dilemma in that it involves a contradiction between ideas and values that are explicitly held" (Bowers, 2011a, p. 38). Bowers offers as an example of double bind thinking: "when consumer-oriented economic growth actually undermines the natural systems we depend upon" (2011a, p. 39). He provides examples of the ways that these conceptual double binds are perpetuated in academic language. Today, however, we are witnessing rejection of the knowledge of accomplished academics in the health care disciplines by Trump and many in his base. This rejection simply draws on and maintains other kinds of destructive root metaphors – *hoax, witch hunt, fake news,* etc.—that cause serious double bind thinking amongst supporters of this president. Trump has so often repeated these particular words in multiple contexts—Russia investigation for conspiracy and obstruction of justice, impeachment hearings, and now the pandemic— that they have become new root metaphors for this time. These terms are levied at experts in government policy, law, journalism, and now public health, undermining the public's ability to discern any sort of truth. No doubt, this is not the sort of criticism of academia and other experts that Bowers would endorse. It seems a clever cooptation and distortion of the academic tradition of skepticism and critique, and is antithetical to Bowers' call for taking a broader, ecologically attuned, view of our language and practice. We cannot know precisely how Bowers might have chosen to respond to this but can reasonably assume that he would have sought to distance his own critique from that of the Trump right.

Solidarity versus Incivility and Anti-Social Behavior

As people fell ill, vulnerable populations became house-bound, and millions lost jobs, the outpouring of help from communities was enormous. People picked up and delivered groceries, donations poured into the food banks. When it became clear that there were not enough masks for front line health care workers, people began sewing them from cloth and donating thousands (Lipner, 2020). Local distilleries began producing hand sanitizer, a commodity that had quickly disappeared from grocery store shelves (Shilton, 2020). Some industries shifted manufacturing to produce other personal protective equipment (PPE) such as gowns, shields, and masks (Camillo, 2020).

The double face of the situation, however, was also present. Many hoarded scarce commodities such as surgical masks, toilet paper, hand sanitizer, and certain grocery items. Some of the hoarding reflected fear of not having enough, but much of it was carried out by opportunists who then sold it back to the public at steeply inflated prices, a rational choice in a culture driven by root metaphors of advanced capitalism. Despite the eventual recommendation by the CDC for masks when in public to reduce transmission of the virus, the president refused to wear one and many of his supporters followed suit. Cashiers who did not have the luxury of working from home, but were needed to supply food, reported daily instances of extreme rudeness from some customers, including some who refused to wear masks that would protect cashiers (Editorial, 2020). Some purposely coughed on food and others, in one case necessitating the destruction of $35,000 worth of food (Fieldstadt, 2020). And, as described already, small bands of protesters surrounded state capitols, some armed, to insist all businesses be allowed to open pre-maturely (pre-mature, that is, according to the CDC). Threats of assassination for the democratic governor of Michigan were posted on social media by one of these groups (Mele, 2020).

Role of the Internet

The anti-social reactions to a crisis that are driven by root metaphors of advanced capitalism—versions of individualism that emphasize greed and selfishness—are reinforced and accelerated by our new digital dependence. "News" that is truly fake goes viral in internet communities when it reinforces both implicit and explicit biases of the target groups. Discerning legitimate from illegitimate sources on the internet can be tricky even for those who attempt to trace sources, impossible for those who make no such attempts or who do not know how. Bowers has expressed deep concerns over many years about the role of digital devices and the internet in our lives. Digital technology, he argued, has threatened cultural commons by making wisdom traditions—those that have insight into how we might extricate ourselves from this ecological quagmire—irrelevant, favoring computer generated data as legitimate knowledge (Bowers, 2000, 2016a). The instant availability of massive troves of decontextualized information can create the illusion that knowledge from past traditions—sewing, cooking, farming, building cooperative communities, morality, for example—is irrelevant. He noted the ways digital technology could diminish our long-term memory now that we can instantly look up anything. It can impair our ability to focus given the ceaseless distractions that pop up in any internet search. These factors weaken our ability to listen to others and to navigate the offline world, and this isolates us and promotes the myth of the autonomous individual. Further, digital technology, notably artificial intelligence, is rapidly eliminating the need for human labor in many sectors, destabilizing communities, promoting greater movement toward unsustainable corporate capitalism, and increasing the threat of violence (Bowers, 2001, 2016a).

It is easy to document the ways Bowers' concerns about digital technology were well founded. We continue to see the violence, both real and symbolic, spawned from the wide and instantaneous distribution of disinformation, the uses and abuses of computer-generated data, and more. The pandemic, however, revealed unexpected beneficence to one of the (double) faces of digital technology—a beneficence that could prove one of the keys to revitalizing the cultural commons. When the decision was made that we must shut down face to face businesses, except for those deemed essential, and isolate in our homes, more than half of all employed Americans were able to work from home via computers and phones (Zapier Editorial Team, 2020). For teachers at all levels, this meant doing the difficult work of rapidly transforming classroom plans to make them suitable for online and/or remote learning—a radical reconceptualization of the work of education. For K/12 schools the general consensus has been that children's education suffered (Hobbs, 2020). In addition to the rushed transition, there were access issues for many students. This was true at the college level as well. However, the majority of those working from home (not teachers) have so far reported that their productivity increased, and they enjoyed seeing family members during the day (Zapier Editorial Team, 2020). Tools such as Zoom enabled remote meetings as well as synchronous classrooms online.

Now that teachers have had more time to learn more about the technology available and how to adapt their teaching to remote and online formats, it seems likely student experience will improve. Through multiple media publications, however, teachers and students tend to report that they much prefer face to face classrooms for the communication capacity those foster. Given the constraints of upcoming return to the classroom—masks, social distance, busing, and more—ironically, online synchronous classes may provide superior opportunities to converse, discuss, and read one another's facial expressions. By the time this book is in the hands of readers we will know more about how this played out.

Internet during the pandemic also enabled people to organize and help one another quickly. With so many of the elderly and disabled stuck at home, getting basic needs met could be a serious problem. Many volunteered to do grocery shopping, pick up prescriptions and other necessities, and deliver to people in need. In my state of Vermont, which is largely rural and divided by mountains, a web service called Front Porch Forum that is freely available to everyone in the state made possible rapid communication about needs for volunteers and donations. (It has been available for these purposes long before the pandemic but proved especially useful at this time.) Though this medium does not provide face-to-face exchanges within a community of the sort Bowers (2016a) advocates, it hijacks the usual trajectory of digital technology and adapts it as a pathway to intergenerational knowledge and care exchange in the cultural commons.

These experiences have enabled many of us to imagine different ways of conceptualizing work. We now understand that we do not need to travel so much in order to do our work and to connect with others. We have had the opportunity to slow down and appreciate our own homes. Bowers anticipated this when he wrote, "[the wealth of the cultural commons] enables people to discover interests and talents that lead to less stressful and thus less debilitating lives, to lifestyles that have a smaller adverse impact on the ability of natural systems to renew themselves" (Bowers, 2012a). Of course, not everyone has had equal opportunity for these kinds of experiences as will be explored shortly. COVID-19 makes clear, if it were not so before, that access to the internet should be a universal right.

Deep Inequities Laid Bare

The cruelty of our consumer-dependent monetized economy has been revealed for more to see. Income inequality in the United States was already at "Gilded Age" levels prior to the pandemic: The top 10 percent have nine times more income as the bottom 90 percent; the top 1 percent have 39 times more; the top 0.1 percent have 196 times more (Institute for Policy Studies, 2020). The pandemic is highlighting and exacerbating those inequalities. We are seeing the stark disparity between those who are able to work from home and those who have been laid off, furloughed, or fired. Forty-four million are unemployed as I write this; many living paycheck to paycheck are now lined up at food banks in unprecedented numbers. Those who can work from home also tend to have access to digital resources for children to do online learning, the leisure to do things like bake bread, exercise outdoors in parks, and other recreational opportunities. Those who have little to no income often have limited access to internet, which also makes it almost impossible for children home from school to participate in online lessons. Many are food insecure. Many "front line workers" such as food industry employees and delivery persons are paid minimum or low wages and expected to risk lives for that. Many have no health insurance or paid sick leave. Lack of universal health care and paid sick leave endangers everyone in a pandemic, something few who argue against these benefits for all have been confronted with in such unambiguous terms. Meanwhile billionaires have grown even more wealthy during the pandemic. Benefitting from COVID-19 "disaster capitalism" (Klein, 2007), Jeff Bezos, CEO of Amazon, grew his wealth by $34.6 billion between mid-March and mid-May 2020; Mark Zuckerberg, CEO of Facebook, added $25 billion during the same period (Frank, 2020). It should be noted, this astronomical gain in wealth is primarily for those in the tech industries.

The inequality gap also puts those at the bottom at greater risk from COVID-19. Lack of health insurance or paid sick leave is only one part of that. Lack of access to healthy food and healthy housing makes one more vulnerable to the disease. Stress compromises the immune system. We are also seeing a racial divide with African Americans at much greater risk from the virus. The ripple effects of economic deprivation, higher risk of becoming infected and of transmitting the virus, lack of quality childcare and more are devastating and will ultimately impact the fabric of society. Bowers warned of the "consequences of these largely ignored realities," that is, expanding inequality, job, food, and health care insecurity that have come from "increased automation, ... outsourcing of jobs to low-wage regions of the world, the breakdown in the social contract between employers and employees ... " (Bowers, 2012b, Ch. 3 ebook).

Those activists and politicians who have fought for social safety nets to mitigate these inequalities, to bring about universal health care, student debt forgiveness, and other expanded benefits have been told by opponents that the United States cannot afford such "entitlements." As the nation watched a panicked legislature address the economic and public health crisis that is COVID-19 with multi-trillion-dollar aid packages, "we cannot afford social safety nets" was revealed as a lie. How much more might have been done had there not been a multi-trillion-dollar tax cut in 2017, for which an outsized share of the benefits goes to corporations and the top 1%.?

Environmental Revelations

In April, there was a 17 percent reduction in global emissions (Mooney, 2020). City skylines in places like Los Angeles, industrial cities in China were seen clearly for the first time in years. Punjab and Nairobi had views of mountains that had been obscured by haze for years. Satellites revealed cleaner air across large areas in Asia, Europe, and North America (Cohan, 2020). Waterways in Venice cleared (Condie, 2020). Most of these changes were attributed to the huge reduction in ground travel. Gasoline prices plummeted for lack of demand. A significant percentage of the reduction was due to reduced air travel (Mooney, 2020). People are discovering that business travel is not necessary, nor is travel to and from work for many, thanks to internet. We have been shown that we can indeed reduce our carbon footprint but that requires some tremendous sacrifices under the current economic system. Climate scientists warn that we must reduce our emissions globally by 7.6 percent every year between 2020 and 2030 in order to limit global temperature rise to 1.5 percent, though this might not be enough. What is clear is the need to restructure an economy so heavily dependent on fossil fuels and consumerism. These phenomena bring home the idea that we need to rethink on how we need to live to preserve the planet for future generations and for all life on earth.

We are in the midst of the sixth mass extinction in large part due to climate change, but also due to our encroachment into animal and plant habitats (Kolbert, 2014). With humans sheltering in place, traveling less, staying indoors, animals have been sighted in unusual places around the world. In Brazil endangered baby sea turtles made their way to the ocean unimpeded by artificial light and other human activities for the first time in more than a decade. Wild boar entered the streets of Barcelona. Mountain goats wandered through a town in Wales. Illegal wildlife trade has been curtailed. A mountain lion was spotted lounging in a tree in downtown Boulder, Colorado (McCoy, 2020). These are temporary respites for the animals, but their presence offers another revelation to humans. We need to give other creatures a break and some space. Bowers wrote of "[t]he power of mythical thinking to distort awareness of what should be obvious to everyone ... [including] how the exploitation of the natural environment is not leading to progress, but to greater scarcity and impoverishment" (Bowers, 2016a, p. 43). The pandemic has offered the unique opportunity, in this and many other ways I have attempted to lay out in this essay, to unveil this deeply embedded mythical thinking for what it is. It has given us a glimpse in real time of a revitalized environmental commons.

Revelations from Sheltering in Place

The forced slowdown of staying at home, it turns out, was something many of us needed whether we knew it or not. The pace of our work in support of a monetized economy interferes with reflection and has limited our ability to see beyond the root metaphors that we have so deeply internalized. We move too fast through our days and our activities largely discourage contemplation about meaning. On the one hand, the luckiest and most privileged among us were able to reflect about the quiet and calm of staying home, shopping only for groceries, a simplified life that offered time for real reflection about what matters. The lockdown interrupted consumerist behaviors and illuminated the value of one's place. It also illuminated the losses at a time when we had the space in which to grieve. In short, it gave us a window into what a revitalization of the cultural commons might do to bring psychological and physical healing, just as Bowers has suggested it could. This experience has the potential to grow the resistance to what Bowers has called a "perfect storm" of destructive forces by heightening and expanding awareness of the importance of the cultural commons (2016a, pp. 115–116). Posts on my social media page have included reflections on the virtues of a simpler life, less consumerism, less travel, and a heightened awareness of the natural world. The "comments" section after an article in *The Guardian*, "A letter to my post-lockdown self: 'Keep listening to the birds'" (Graham, 2020) revealed similar sentiments. Some noted their intent to attend more to loved ones and to work differently after the pandemic. Others committed to doing more for those who suffer the most in our consumeristic society.

On the other hand, some were forced into abusive situations, exacerbated by tensions of job loss, anxiety about survival, bad relationships. Some faced poverty, debt, and hunger. Some were stressed by the demands of working at home at the same time they must help children with school, and they must care for children all day. And some were deemed "essential workers" having to face the public daily when the public was often at its worst. These circumstances have also been illuminating—especially for those who experienced them, but also for those who have been witnessing, paying attention to the plight of our communities. The damage done to human beings by economies based on growth, consumerism, greed, and austerity for some (the opposite for the wealthiest) has been addressed through much of Bowers' work, particularly in his description of "enclosure": "the process by which something that is shared in common is turned into something that is privately owned, then monetized and integrated into the industrial/market economy" (2016a, p. 116). Now previously denied inequities have been laid bare, made obvious.

In the Wake of a Pandemic; How Will We Be Changed?

The pandemic and lockdown revealed little about our economic and cultural systems that were previously unknown. Many journalists, scientists, and social theorists, including Chet Bowers, have been writing about these issues for years and even decades. However, the pandemic has become a kind of laboratory where more people are actually able to *experience* the effects of inequality, food insecurity, scarcity, environmental degradation and what it might mean to reverse that, and our dependence on experts and one another.

While some citizens wanted to re-open the economy and lift the lockdown sooner than CDC said was prudent (in late April or early May), polls showed that most Americans wanted to play it safe. Nevertheless, there was pressure from business and the White House for states to begin reopening early. That pressure coincided with anti-lockdown protests that included "anti-vaccine activists and other conspiracy theorists, rightwing provocateurs, members of known anti-government militias, gun rights advocates, established conservative groups backed by wealthy billionaire donors, Republican stalwarts and people who were actually out of work" (Beckett, 2020). Historian Roxanne Dunbar-Ortiz illuminated a stunning example of double bind thinking when she warned, "The capitalist class, those who benefit most from the unequal system, they know it's not sustainable. They're desperate not to stay locked down too long, so people get used to fresh air, breathing air without carbon in it. People might get ideas of a different kind of world" (Beckett, 2020).

What might that different kind of world be? As I write in late July, 21 states that reopened early are seeing a resurgence in cases, hospitalizations, and deaths. The virus will likely be with us for months and maybe years to come. The timeline for a vaccine is uncertain, and when there is one, its efficacy and duration will also be uncertain. It is reasonable to expect that our behaviors around one another will change in response to the threat of contagion. But what will come of our newfound understandings of the environmental, economic, and cultural systems under which we live? That remains to be seen but the experience has certainly presented a unique opportunity.

Our educational institutions should not squander this opportunity for deep change. In addressing the task of those in leadership roles, Bowers presented "key ideas and assumptions that will lead to more ecologically sustainable cultures" (Bowers, 2011b). Any educator, whether in a leadership role or not, could apply these ideas to curriculum and instruction. The pandemic experience, having offered insight through experience, is an excellent place to start.

With shared stories of lockdown and other pandemic experiences, students (and teachers) might be invited to reflect upon the "differences in their experience when participating in activities of the cultural commons and when participating in the relationships and activities that are part of the consumer/monetized culture" (Bowers, 2009). In what ways did we learn about our interconnectedness to others and to the natural world? How has our understanding of certain root metaphors such as efficiency (versus, for example, resilience), equality, democracy, changed? How might our ecological intelligence have been changed or cultivated by slowing the pace of life, having time to observe, and noticing the changes in our environment? What are some of the traditional values we inherited or learned recently that enriched our lives during the pandemic, and what are some that were revealed to be flawed? What happens to the cultural commons when some members are denied such rights as paid sick leave and health care, or safe workplaces? How did the lockdown affect our attitudes toward and relationship with material goods? How are we to understand the root metaphor "freedom"—for example, freedom to not wear a mask versus freedom to have public health protected? What histories might help us understand all of these experiences more fully?

Notes

1. Chet Bowers suggested that the erosion of the quest for wisdom—a quest that was central to indigenous cultures of the world—has been colonized by "followers of the scientific method" (Bowers C. A., 2018, p. 200). It would seem that the astonishing scientific and technological achievements of recent centuries have come to mean that "we do not need to understand wisdom" (p. 200), including ecological wisdom that came from thousands of years of observation and survival in particular ecosystems.
2. A similar collision between radical right populist leaders and large countries where the pandemic has been mishandled can be seen also in Russia, Brazil, and Great Britain (Leonhardt, 2020).
3. Foer makes the argument in his book, *We Are the Weather*, that animal agriculture is responsible for 51 percent of global greenhouse gases, which is what the Worldwatch Institute claims. Other estimates are significantly lower at around 14.5 percent (still large!). An appendix in Foer's book plays out the differences in the studies and the reasoning behind them, and ultimately the reason he believes that 51 percent is the more accurate estimate.

References

Anoruo, N. (2020, May 14). How social media undermined the COVID-19 response. *ABC News.* Retrieved from: https://abcnews.go.com/Health/social-media-undermined-covid-19-response/story?id=70511613

Beckett, L. (2020, May 21). 'All the psychoses of US history': How America is victim-blaming the coronavirus dead. *The Guardian.* Retrieved from: https://www.theguardian.com/world/2020/may/21/all-the-psychoses-of-us-history-how-america-is-victim-blaming-the-coronavirus-dead

Bowers, C. A. (2000). *Let them eat data: How computers affect education, cultural diversity, and the prospects of ecological sustainability.* Athens, GA: University of Georgia Press.

Bowers, C. A. (2001). *Educating for eco-justice and community.* Athens, GA: University of Georgia Press.

Bowers, C. A. (2010). Understanding the connections between double bind thinking and the ecological crises: Implications for educational reform. *Journal of the American Association for the Advancement of Curriculum Studies, 6*(1), 1–14.

Bowers, C. A. (2011a). *University reform in an era of global warming.* Eugene, OR: Eco-Justice Press.

Bowers, C. A. (2011b). The leadership role of University Presidents, Deans, and Department Chairpersons. In C. Bowers (Ed.), *University reform in an era of global warming* (pp. 127–128). Eugene, OR: Eco-Justice Press, LLC.

Bowers, C. A. (2011c). *Educational reforms for the 21st century: How to introduce ecologically sustainable reforms in teacher education and curriculum studies.* Eugene, OR: Eco-Justice Press.

Bowers, C. A. (2012a). The political economy of the cultural commons and the nature of sustainable wealth. In C. A. Bowers (Ed.), *The way forward: Educational reforms that address the cultural commons and the linguistic roots of the ecological/cultural crises.* Eugene, OR: Eco-Justice Press.

Bowers, C. A. (2012b). *The way forward: Educational reforms that address the cultural commons and the linguistic roots of the ecological/cultural crises.* Eugene, OR: Eco-Justice Press.

Bowers, C. (2016a). *Reforming higher education in an era of ecological crisis and growing digital insecurity.* Anoka, MN: Process Century Press.

Bowers, C. A. (2016b). How technological forces and the ecological crisis are leading to a more violent future. In C. A. Bowers (Ed.), *Reforming Higher education in an era of ecological crisis and growing digital insecurity* (pp. 43–72). Anoka, MN: Process Century Press.

Bowers, C. A. (2017, January 21). *cabowers.net.* Retrieved from: https://cabowers.net/dicterm/CAdict024.php

Bowers, C. A. (2018). *Ideological, cultural, and linguistic roots of educational reforms to address the ecological crisis: Selected works of C. A. Bowers.* New York, NY: Routledge.

Camillo, J. (2020, June 17). Manufacturers shift to PPE production to fight COVID-19 pandemic. *Assembly Magazine.* Retrieved from: https://www.assemblymag.com/articles/95741-manufacturers-shift-to-ppe-production-to-fight-covid-19-pandemic

Cohan, D. (2020, April 17). COVID-19 shutdowns are clearing the air, but pollution will return as economies reopen. *The Conversation.* Retrieved from: https://theconversation.com/covid-19-shutdowns-are-clearing-the-air-but-pollution-will-return-as-economies-reopen-134610

Condie, S.. (2020, April 5). Coronavirus lockdowns clear the air, but the green effect could be fleeting. *The Wall Street Journal.* Retrieved from: https://www.wsj.com/articles/coronavirus-lockdowns-clear-the-air-but-the-green-effect-could-be-fleeting-11586095204

Editorial. (2020, May 13). Cashiers shouldn't have to put up with rude behavior. *Milford Daily News.* Retrieved from: https://www.milforddailynews.com/opinion/20200513/our-view-cashiers-shouldnt-have-to-put-up-with-rude-behavior

Fieldstadt, E. (2020, March 26). Woman who coughed on $35K worth of grocery store food charged with four felonies. *NBC News.* Retrieved from: https://www.nbcnews.com/news/us-news/grocery-store-throws-out-35k-worth-food-woman-coughed-twisted-n1169401

Foer, J. S. (2019). *We are the weather: Saving the planet begins at breakfast.* New York: Farrar, Straus, Giroux.

Frank, R. (2020, May 21). American billionaires got $434 billion richer during the pandemic. *CNBC.* Retrieved from: https://www.cnbc.com/2020/05/21/american-billionaires-got-434-billion-richer-during-the-pandemic.html

Graham, J. E. (2020, May 23). A letter to my post-lockdown self: 'Keep listening to the birds'. *The Guardian.* Retrieved from: https://www.theguardian.com/lifeandstyle/2020/may/23/a-letter-to-my-post-lockdown-self-keep-listening-to-the-birds#comment-140921411

Hobbs, T. (2020, June 5). The results are in for remote learning: It didn't work. *The Wall Street Journal.* Retrieved from: https://www.wsj.com/articles/schools-coronavirus-remote-learning-lockdown-tech-11591375078

Institute for Policy Studies. (2020). Income Inequality in the United States. *Inequality.org.* Retrieved from: https://inequality.org/facts/income-inequality/

Klein, N. (2007). *The shock doctrine: The rise of disaster capitalism.* New York: Picador.

Koch, A, Bierley, C., Maslin, M., and Lewis, S. (2019, January 31). "The Conversation" in the world. Public Radio International. Retrieved from: https://www.pri.org/stories/2019-01-31/european-colonization-americas-killed-10-percent-world-population-and-caused

Kolbert, E. (2014). *The sixth extinction: An unnatural history.* New York: Henry Holt and Company.

Leonhardt, D. (2020, June 4). Column: Virus is growing most in countries with 'illiberal populist' leaders. *Chicago Tribune.* Retrieved June 2020, from: https://www.chicagotribune.com/coronavirus/sns-nyt-virus-growing-most-countries-illiberal-populist-leaders-20200604-d65itucnx5htdhdptumryy3hqy-story.html

Lipner, M. (2020, March 31). Volunteers work with hospitals to make emergency face masks for workers. *Today.* Retrieved from: https://www.today.com/health/volunteers-sew-homemade-face-masks-hospital-workers-t177079

McCoy, T. (2020, April 15). As coronavirus sends humans indoors, wild animals take back what was once theirs. *The Philadephia Inquirer.* Retrieved from: https://www.inquirer.com/health/coronavirus/coronavirus-covid-19-wildlife-impact-social-distancing-20200415.html

Mele, C. (2020, May 15). Man faces terrorism charge after threatening to kill Michigan's Governor, officials say. *The New York Times.* Retrieved from: https://www.nytimes.com/2020/05/15/us/virus-michigan-whitmer-threats.html

Mooney, C. D. (2020, May 19). Global emissions plunged an unprecedented 17 percent during the coronavirus pandemic. *The Washington Post.* Retrieved from: https://www.washingtonpost.com/climate-environment/2020/05/19/greenhouse-emissions-coronavirus/?arc404=true

Packer, G. (2020, June). Underlying conditions. The Atlantic, Washington, D.C.

Peeples, L. (2020, May 18). To prevent pandemics, bridging the human and animal health divide. *Undark.* Retrieved from: https://undark.org/2020/05/18/human-animal-medicine-pandemic/

Pollan, M. (2020, June 11). The sickness in our food supply. *The New York Review of Books.* Retrieved from: https://www.nybooks.com/articles/2020/06/11/covid-19-sickness-food-supply/

Samuel, S. (2020, May 12). Our environmental practices make pandemics like the coronavirus more likely. *Vox.* Retrieved from: https://www.vox.com/future-perfect/2020/3/31/21199917/coronavirus-covid-19-animals-pandemic-environment-climate-biodiversity

Samuel, S. (2020, June 10). The meat we eat is a pandemic risk, too. *Vox.* Retrieved from: https://www.vox.com/future-perfect/2020/4/22/21228158/coronavirus-pandemic-risk-factory-farming-meat

Shilton, A. (2020, April 17). This Vermont distillery is now making hand sanitizer. *Outside.* Retrieved from: https://www.outsideonline.com/2411675/caledonia-spirirts-vermont-distillery-hand-sanitizer

Stevens, H. (2020, March 31). From 'It's going to disappear' to 'WE WILL WIN THIS WAR'. Washington Post.

Zapier Editorial Team. (2020, April 6). Half of America just started working from home. So, how's it going? *Zapier.* Retrieved from: https://zapier.com/blog/wfh-report/

Part III

Ecojustice Curriculum

According to Brown, root metaphors are the "meta-schemata" that frame the process of analogic thinking across a wide range of cultural experience (1978, p. 126). They are often based on the mythopoetic narratives of a cultural group, or on powerfully evocative experiences. Patriarchy and anthropocentrism are root metaphors that can be traced back to the Biblical account of creation.

Mechanism is a root metaphor that had its roots in the transition from a Medieval to the modern world view. Other meta-schemata or root metaphors underlying modernity include linear progress, evolution, economism, and the autonomous individual. The way in which root metaphors carry forward earlier culturally specific patterns of thinking can be seen in how the root metaphor of mechanism, which Johannes Kepler used to explain the universe as a "celestial machine," continues to influence the current fields of medicine, architecture, education, brain research, and even the Human Genome Project. The difficulty in recognizing how thought and even the material expressions of culture are based on root metaphors that reproduce past forms of cultural intelligence and moral norms can be seen in how the supposedly most rationally capable members of society were unable to recognize the many expressions of patriarchy within the relationships and curricula of the university.

When root metaphors are not recognized they largely dictate which analogs will be used to understand new phenomena. Thus, the process of analogic thinking—that is, fitting the new into the old schema, to recall Nietzsche's observation—reproduces the older conceptual patterns. When there are several competing ways of understanding a new phenomenon, the analogy that prevails becomes over time an iconic metaphor that is a taken-for-granted aspect of thought and communication. Examples of iconic metaphors include "data," the personal pronoun "I," "emancipation," "freedom," "equality," "domination," "environment," and so forth. Indeed, most of our thought and discourse, even material expressions of culture, are dependent upon the use of iconic metaphors that reproduce the analogies that prevailed at an earlier time due to the dominant status of a root metaphor. When there are competing root metaphors, such as between "ecology" and the collection of root metaphors underlying the Industrial Revolution, iconic metaphors such as "sustainability" have different meanings that reflect the differences in taken-for-granted root metaphors. The constitutive role of root metaphors in framing thought can be summarized by paraphrasing Nietzsche and Heidegger in the following way: Language thinks us as we think within the conceptual categories that the language of our cultural group makes available. As thought is inherently metaphorical, there is always the possibility of identifying more adequate analogies, and even of recognizing aspects of cultural/personal experience that previously held root metaphors cannot account for.

In order to recognize the profound difference in the root metaphors that underlie an ecojustice pedagogy, it is first necessary to identify the root metaphors that are taken-for-granted by critical pedagogy theorists. Their root metaphors, as I suggested earlier, were the basis of the Industrial Revolution that is being continued today under the new metaphor of "globalization," which has replaced the older and highly criticized metaphor of "colonization," Contrary to their claims, the practice of a critical pedagogy does not lead to individual emancipation and social justice; rather it reinforces a subjectively centered individualism required by the consumer, technologically dependent society. While they are highly critical of the capitalist foundations of society, the root metaphors that underlie their prescriptions for change create a double bind they fail to recognize (pp. 22–23).

The Root Metaphor Underlying an Ecojustice Pedagogy

The rapid decline in the viability of natural systems, along with the current rush to globalize the western consumer lifestyle, is already introducing environmental changes that bring into question the assumptions upon which the western mind-set is based. Indeed, the language based on the assumptions about progress, a human centered world, and individualism leads to such a distorted understanding that environmentally caused diseases, cleaning up oil spills, and efforts to reverse degraded ecosystems are at least in North America, treated as economic activities that contribute to the gross domestic product. More important is the way in which earlier assumptions encoded in the metaphorical language lead to pursuing the very policies and developments that further exacerbate the crisis. The policies of the World Trade Organization and the scientific efforts to further industrialize agriculture are two prime examples of the double bind associated with the language of progress.

As feminists discovered, changing the root metaphor of patriarchy was a long and difficult process– one that is still underway. The beginnings of a shift in root metaphors is now taking place within the environmental sciences, among a few heads of corporations who are beginning to realize that production processes must mirror the design patterns found in nature, and among theologians who are attempting to find scriptural authority for an environmental ethic. Ecology, the emerging root metaphor, can be traced back to the Greek word "oikos" which referred to the maintenance of relationships within the family household. Without going into the history of how the original analog was transformed into the scientific study of relationships within natural systems, I want to point out several reasons this metaphor can be expanded in ways that clarifies and legitimates an ecojustice pedagogy. The use of ecology as a root metaphor (which means it should guide the conceptualization of the widest possible range of cultural practices) foregrounds the relational and interdependent nature of our existence as cultural and biological beings. This includes our participation in a highly complex web of symbolic relationships deeply rooted in the past. We could neither think nor communicate if we were isolated from the language systems that sustain the patterns of cultural life, and which are the basis for their gradual transformation.

Our participation in the even more complex web of interacting systems that constitute the natural world involves a more basic form of dependency. The oxygen we breath, the sources of nourishment, and even the autopoietic networks that interact at the genetic level to create the living system we know as our biological self, are interconnected across many scales of life producing systems. These cultural and biological processes lead to biographically distinct expressions of individualization–which we consider our self-concept, and conceptual and moral proclivities. These processes, which Humberto Maturana and Francisco Varela refer to as the "structural coupling of autopoietic systems"(1987, pp. 75–80), can also be understood as ecological systems— which serves as the root metaphor that foregrounds relationships, continuities, non-linear patterns of change, and a basic design principle of Nature that favors diversity.

An educational process based on this root metaphor must recognize that living systems involve both the replication (conservation) of patterns of organization as well as changes introduced by internal and external perturbations. It also needs to recognize the importance of diverse cultural systems that develop in response to the differences in natural systems. Some cultures failed to adapt to changes in natural systems (or introduced environmentally destructive changes) and thus reduced their own chances of survival. Other cultures, however, have become repositories of knowledge of local plants, animals, and natural cycles that affect their sources of food, water, and other cultural necessities. In effect, this root metaphor foregrounds the different forms of interdependencies, as well as the need to exercise critical thought in ways that strengthen the ability of natural and cultural systems to renew themselves in ways that do not compromise the prospects of future generations.

When based on the root metaphor of an ecology, an ecojustice pedagogy has three main foci: (1) Environmental Racism and Class Discrimination. The disproportionate impact of toxic chemicals on the health of economically and ethnically marginalized groups is part of a cycle that encompasses more than the political process that determines where toxic waste sites and industries are to be located. In addition to class and racial biases, the cycle includes the phenomenal growth in the last 50 years in the use of synthetic chemicals (estimated at over 80,000), and a level of personal consumption based on rising rates of resource extraction and manufacturing—all of which have created monumental waste disposal problems. In short, the consumer/technology dependent lifestyle in the West, which is now being promoted in "undeveloped" regions of the world, increases the impact of contaminated environments on those groups least able to protect themselves.

An ecojustice oriented education needs to inform students about the politics of toxic waste disposal, which not only encompasses minority and working class communities but also crosses national boundaries in ways that spreads misery to Third World countries. Students need to learn how different groups are resisting the contamination of their local environments and workplace, and how the politics of environmental discrimination works. Attention should also be focused on the contaminated work environments that pose special health risks for poor workers in Third World countries where many of the goods consumed in the West are produced. (2) Recovery of the Non-Commodified Aspects of Community. According to Paul Hawkin, and Amory and L. Hunter Lovins, "industry moves, mines, extracts, shovels, burns, wastes, pumps, and disposes of 4 million pounds of material in order to meet one average middle-class American family's needs for a year" (1999, p. 50). This amount of material flow is slightly less for Canadians, and may vary by half in other Western countries. The shared trendline, however, is toward increasing dependence on meeting life's daily needs through consumerism rather than through self-reliance within the family and networks of mutual support within communities. The critical distinction, though not always clear cut, is between the growth of commoditized knowledge, skills, and relationships which the industrialized system (even in the era of e-commerce) requires, and what remains of the non-commoditized individual, family, and community patterns of daily life. The relentless drive to commoditize more aspects of daily life, and thus to create new markets and thus new forms of dependencies, is a key factor in the cycle of production, product obsolescence and misuse, and environmental contamination that is contributing to the rapid changes we are witnessing in natural systems.

The implications for an ecojustice pedagogy include providing a critical understanding of the deep cultural assumptions that underlie the industrial and consumer dependent form of culture as well as an understanding of how the languaging patterns of different western cultures create the individual psychology that accepts consumer dependency and environmental degradation as a necessary trade-off for achieving personal conveniences and material success. But more is required of an ecojustice pedagogy than the development of critical awareness. There is a constructive side as well. Learning the principles of ecological design, and how they can be applied to buildings and technologies in the students' bioregion is critically important to moving away from the industrial model that still prevails. There is also a need to use the educational process to regenerate the non-commoditized skills, knowledge, and relationships that enables individuals, families, and communities to be more self- reliant—and thus to have a smaller ecological impact. Many readers might interpret this suggestion as the expression of a romantic nostalgia for earlier lifestyles that were actually characterized by poverty, debilitating hard work and shortened lives. Rather, what is being proposed as a way of reaching a better balance between self-sufficiency and consumerism (perhaps even reversing the degree of consumerism) is for a curriculum that helps students recognize the extent their daily lives depend upon commoditized relationships and activities. The curriculum should also help them recognize the patterns and activities within their own communities that are still largely based on face-to-face, intergenerational sharing of knowledge and skills. These non-commoditized aspects of family and community life might range from dinner conversations made possible by a more balanced use of such modern technologies as television and computers to the existence of community theater and other performing arts, mentoring in the development of individual talents, gardening, chess and poetry clubs, sports, and community service activities.

Learning about (and thus valorizing) the non-commoditized traditions of ethnic minorities should also be part of an ecojustice curriculum. Many of these cultural groups have survived economically and politically repressive environments because of their ability to carry forward the intergenerational knowledge that enabled them to be less dependent upon the consumerism that more privileged groups took for granted. In suggesting that marginalized cultural groups still retain non-commoditized traditions that need to be reinforced, rather than be undermined by the emancipatory approach of critical pedagogy theorists, I am not suggesting that the curriculum should reinforce the inequitable patterns that keep some cultural groups living below the poverty line and in degraded environments that create greater health risks. Rather, the educational challenge is to contribute to their having more equal access to educational opportunity, political empowerment, and an improved material standard of living. What needs to be avoided is exposure to a curriculum that denigrates their heritage of intergenerational knowledge–which may include elder knowledge, patterns of mutual aid and solidarity that link together extended families and community networks, ceremonies, narratives, and other traditions essential to their self-identity and moral codes. (3) Responsibility to Future Generations. The prospects of future generations should also be a central focus of an ecojustice pedagogy (pp. 28–31).

An ecojustice pedagogy contributes to self-limitation for the sake of future generations when it helps students recognize and participate in the non-commoditized activities of community. But it is not a form of self-limitation that undermines the student's well-being; rather it represents an expansion of relationships and opportunities to develop personal talents that can further enrich the community. An ecojustice curriculum that helps introduce students to the non-commoditized possibilities of community is only part of the reform that is needed. The prospects of future generations are also dependent upon today's students acquiring the conceptual basis for democratizing technology and science. The root metaphors that gave conceptual direction and moral legitimacy to the Industrial Revolution continue to frame the public's understanding of technology and science—which they view as the highest embodiment of progress. While modern technology and science have made many genuine contributions to improving the quality of life, the fact remains that they provide the basis for extending the industrial model into all areas of food production (which is leading to the narrowing of the genetic basis of the world's food supply), human reproduction, thought and communication, and education.

As technologies are altering the chemistry of life and changing cultural patterns in ways that are not predictable, the need for basing decisions on the public interest rather than the career interests and profit motive of the elites who create them, should be the paramount concern. The extrapolations of scientific discoveries and theories are also introducing changes in the symbolic foundations of different cultures, with consequences that are even more difficult to predict. Denmark, in particular, has shown leadership in democratizing technology and science that has direct curricular implications, and should ease the fear of educators that democratizing technology and science will undermine the autonomy traditionally associated with these fields of endeavor. But the democratizing process needs to be guided by an intergenerational perspective that takes into account future generations–which teachers need to help students understand.

More specifically, students should learn about the differences between traditional and modern technologies, how modern technologies influence our language and thought patterns, how industrial approaches to technology have transformed communities and deskilled the worker, and how technologies can incorporate the principles of ecological design mentioned earlier (Van Der Ryn & Cowen, 1996; Hawkin et al., 1999). Similarly, the history of western science, the cultural assumptions that underlie its current foci, and its impact as an ideology on the moral foundations of western and non-western cultures, need to be part of the students' understanding if they are to exercise communicative competence in what will be a highly contested discourse. Developments in technology and science are currently guided by the interests of elite groups who have little understanding of the culturally transforming effects–and even less concern with how the cultural changes their discoveries precipitate will impact the quality of life of future generations (p. 32).

*Excerpt for Part III from: Bowers, C. A. (2002). Toward an eco-justice pedagogy. *Environmental Education Research*, 8(1), 21–34.

9   Developing Ecological Literacy as a Habit of Mind in Teacher Education through Ecojustice Progressive Curricula

Kelly Young

The impetus for this chapter comes from my deep gratitude for Chet Bowers' mentorship in the area of ecojustice education (EJE). I have been fortunate to spend time and learn from Bowers and his seminal written work. My intellectual journey began in the fall of 2004 when Bowers gave a lecture at Trent University. Trent University is a small rural academy situated on the Otonabee River in Peterborough, Ontario, Canada. I have been a Professor of Education at Trent University's School of Education since 2003 when our consecutive teacher education program began. It was an exciting time as faculty from the School of Education and Indigenous Environmental Studies came together to learn about Bower's ecojustice framework. I subsequently accepted Bowers' offer to attend the *EcoJustice Education Retreat: Revitalizing Detroit's Commons* in October of that year—a retreat that was organized and sponsored by the Department of Social Foundations at Eastern Michigan University. It was during this time that I met ecojustice-oriented colleagues and I committed to infusing my pre-service curricula with Bowers' (2002) framework as I responded to his call for educational reform in teacher education.

In this chapter, I trace my intellectual journey with Bowers and I outline the ways in which Bowers' ecojustice foundational framework has influenced my department of pre-service teacher education, as well as several pre-service courses, by outlining curricula that promotes Bowers' advocacy for educational reform in terms of reducing "the impact of the industrial/consumer dependent culture on everyday life" (www.bowers. net/ecojusticedictionary). (See also Bowers, 2001, 2002, 2006, 2011; Martusewicz, Edmundson, & Lupinacci, 2011, 2015.) Bowers' understanding of the role of language processes and culture in our complex relationships with the natural world help to lay the foundation of my ongoing curricular infusion of ecojustice principles into pedagogical practices envisioned for the advancement of the field of curriculum studies.

The 2004 *EcoJustice Retreat* inspired my own development of ecojustice as a habit of mind in teacher education. By returning to an excerpt of a previously published article in the currently obsolete *Ecojustice Education*

Review: Educating for the Commons journal, I highlight how Bowers' mentorship and advocacy for ecologically sustainable higher education reform immediately began to influence change in my teacher education curricula (Young, 2005). The following excerpt from my article titled *Developing Ecological Literacy as a Habit of Mind in Teacher Education* documents the importance of ecojustice retreats such as *Revitalizing Detroit's Commons* in 2004 and some of my subsequent infusion of EJE in my pre-service courses:

EcoJustice Education Retreat: Revitalizing Detroit's Commons

In October of 2004, I had the extraordinary experience of participating in an *EcoJustice Education Retreat: Revitalizing Detroit's Commons*, sponsored by the Department of Social Foundations, Eastern Michigan University. The conference began with a bus tour of the restoration of *the commons* throughout the greater Detroit area.

> The commons represent both the naturals systems (water, air, soil, forests, oceans, etc.) and the cultural patterns and traditions (intergenerational knowledge ranging from growing and preparing food, medicinal practices, arts, crafts, ceremonies, etc.) that are shared without cost by all members of the community; nature of the commons varies in terms of different cultures and bioregions; what has not been transformed into market relationships; the basis of mutual support systems and local Democracy; in the modern world the commons may be managed and thus kept from becoming enclosed through private and corporate ownership by being managed by local and national government—municipal water systems and state and national parks are contemporary examples of the commons.
>
> (EcoJustice Dictionary)

Along the tour, we visited the Catherine Ferguson Academy located five minutes from the downtown core of Detroit, Michigan. The school caters to at-risk schoolgirls through an ecological curriculum that integrates the rebuilding and working of a farm as part of its program. All of the core subjects, English, math, science, and social studies involve caring for the farm animals. The girls participated in the building of a barn that is used to house some of the animals. The school incorporates growing plants and vegetables into the curriculum bringing the commons into focus. Our group is told that the families of the schoolgirls became interested in the ecological emphasis of the curriculum and that they were starting to pay attention to the importance of local food production. An emphasis in the program is building conversation about corporate packaged food versus natural food as part of the school's educational project. The students at Catherine Ferguson Academy were being encouraged to see themselves as part of a web of relations and were valuing the commons through mentored relationships from their teachers.

We also toured a local market run by the Coalition of Black Farmers, and a community-shared greenhouse. Our group had the opportunity to experience the grass roots movement in Detroit that involves, in part, taking back the commons for sustainability by communities working together to turn abandoned land into crop producing commons. In Detroit, the grass roots movement involves ecologically progressive projects that were created in response to the needs of economically devastated inner-city communities that are being enclosed from a global commons, of air, water, plants, fish, trees, among others that all humans share. For example, the Detroit local commons is moving further and further toward Enclosure with the privatization of water that is based on an ideological economic realm, in what Vandana Shiva (1993) describes as Globalization toward monotization, which leads to monocultures. Privatization for profit draws upon an ideology that expands enclosure on the commons on local and global scales whereby local regions such as Detroit face enforced subsistence as a result of an industrial consumer mindset.

Later the same day at the conference center, I listened to teacher educators telling stories about how they encountered resistance from teacher candidates toward the development of ecological habits of mind. Teacher educators shared their stories about student resistance to the idea that we all share in a problem of the commons. Many of the stories were familiar to me because it was my experience that teacher candidates simply wanted to know what deep-rooted metaphors and ecological habits of mind have to do with lesson plans and assessment, which is, from their standpoint, the "real work" of teaching. At the conference, I was encouraged to use progressive teaching materials in my course that promote healthy ways of knowing and being by raising the questions, "How do we respond to the problems of the industrial society in post-industrial society? What do we need to conserve? And, what might the pedagogical work of developing ecological habits of mind look like"? What follows is a story of how I incorporated an ecojustice progressive curriculum into my teacher education course on adolescent development.

Ecojustice Progressive Curriculum

Using a (re)conceptualist approach to curriculum, I engage in a "complicated conversation" about ecojustice issues with my students (Pinar 2004). Together, we surf the digital library of the commons and the ecojustice dictionary to familiarize ourselves with ecojustice discourse. Then we viewed the film *Ancient Futures: Learning from Ladakh* (1993) produced by the International Society for Ecology and Culture (ISEC). The film chronicles the changes that the Tibetan Ladakh community encounters when a Western "progressive" industrial revolution permeates their culture, interactions with the natural world and local economy. The (ISEC) writes,

> In the early 1970s, came "development." The first roads into the area brought heavily subsidized goods as well as idealized images of western consumer culture – undermining the local economy and eroding culture self-esteem. The result has been increasing community and family breakdown, unemployment, sprawling urban slums and pollution,
>
> (www.isec.org.uk/ladakh Accessed September 12, 2005)

The self-sustaining and adaptable Ladakh community that had thrived for years on a local-ecological and non-industrial model of community was now experiencing shortages of food that they came to depend upon due to the westernization of ways of knowing and being. As a follow-up activity, I ask my students to think about how in the film the Ladakh people use focal practices to create community (Borgmann, 1992). Focal practices involve building interpretive relationships with the natural world through among other things, gardening. I then ask my students to think about the kinds of focal practices that they themselves engage in with the natural world and compare them to the kinds of focal practices engaged in by the Ladakh community.

By exploring the focal practices of a pre-modern sustainable society such as Ladakh, it becomes clear that "modernity" brings, among other things, the building of roads, which has a serious impact on culture. The film *Ancient Futures* then, provides a microcosm of what has happened in terms of our human place in the story of evolution. It provokes students to make connections back to ancestral thinking prior to conceptually separating cities with wilderness in the imagination (Cobb 1977). It provides a map for thinking about how our health and survival depends upon the natural world because it asks viewers to compare western consumer culture with indigenous sustainable culture. Finally, the film provides insight into the dynamics of the pre-modern world prior to the agricultural revolution and patriarchal structures that are prevalent in our modern world. For example, students are encouraged to consider the ways in which the Ladakh community represents a gender equal society that is not built upon concepts of patriarchy and land ownership.

In addition to sharing my story of my experience in Detroit with my students, I ask teacher candidates to consider their own relationship to the commons by writing their environmental autobiography that is used to help them consider the ways in which the ecology has and continues to play a role in their habits of mind through the question, "What is my relationship to the natural world"? I first encountered the concept of the environmental autobiography in 1998 from Professor Joseph Sheridan at York University in a graduate course titled, *Themes in Storytelling and First Nations Tradition*. By linking my environmental autobiography with a broader understanding of the interrelatedness and co-implication of societies and their relations to natural systems, I am better able to conceptualize ecological literacy is a habit of mind involves connections between people and natural world through interdependence, co-implicated relationships and the valuing of life as sacred (Berry, 1995).

> By sharing my own environmental autobiography, I ask questions that provoke students to consider the ways in which subjectivity is relational and perceptual in relation to the natural world. Taking an interdisciplinary approach to human development and its exceptionalities, I ask my students to engage in an inquiry into the complex relationships of learning and teaching with an emphasis on how various technologies of language and literacy use influence the ongoing development of human senses of self-identity, cultural identity, and ecological identity. My course aims to provide a general overview of cross-disciplinary theoretical developments in human learning (including historical and contemporary perspectives) that have influenced and continue to influence processes, practices and experiences of learning and teaching in everyday institutional and community life by critiquing the taken-for-granted psychological approaches to human development of, among others, Maslow's "hierarchy of needs."
>
> (Excerpt, Young, 2005)

The 2004 EJE retreat clearly influenced my teaching of pre-service teachers. For example, we turned Maslow's hierarchy upside down wherein human physiological needs are paramount as these are directly connected to our local ecosystems. From my experience in 2004 at the EJE retreat, I continued to ruminate upon the ways in which I could infuse teacher education with EJE practices. At this same time, I began attending the *Elders Gathering* that was held annually in February at Trent University. The gathering brought together Indigenous and non-Indigenous learners for several days where traditional teachings and ceremonies were offered by local Elders in the university and in the tipi. Many EJ educators attended each year to learn about Indigenous Knowledge (IKs) through oral storytelling.

In 2006, I attended a second EJE retreat, the *Eco-justice and Educational Reform* retreat in Bowers Bay, Michigan. Each retreat offered many perspectives on the possibilities of EJE. From these experiences, I wanted to bring EJE to a wider audience at Trent University. It was at this time, in 2007, that I coordinated and administered the *International Eco-Justice Education Conference* at Trent University in Peterborough, Ontario,

Canada. EJ educators from across North America came together with students to explore Bowers' framework together with traditional teachings from local Anishinaabe and Haudenosaunee Elders.

AERA Pre-Seminars & EJ Conferences

In April of 2008, at the American Educational Research Association (AERA) in New York city, I attended a pre-seminar on *Eco-Justice and Environmental Education* that was sponsored by Division B—Curriculum Studies. In the summer of 2008, I attended a small *eco-justice* retreat with American EJ educators that was hosted by Dr. Andrejs Kulnieks in Barrie's Bay, Ontario, Canada, as we continued the dialogue about Bowers' work and the need for educational reform in higher education. It was during this time that a new section, "Moral and Ecological Perspectives" was added to Division B under Curriculum Studies. There was a subsequent pre-seminar offered during AERA in San Diego in 2009 where Bowers' deep understanding of the interconnectedness of education and the linguistic and cultural roots of the environmental crisis was articulated in profound ways. I had to opportunity to work with Dr. David Flinders and Dr. Rebecca Martusewicz as the section Co-Chair for the conference. During this time, EJ educators continued to attend the annual *Elders Gathering* with me at Trent University. In addition, in 2012, EJ educators were also offered the opportunity to attend and present at an annual conference on *EcoJustice and Activism* at Eastern Michigan University organized by Dr. Rebecca Martusewicz with the help of Dr. John Lupinacci that brought together EJ educators, researchers, students, and community elders in conversation about Bowers' framework. These annual conferences continued to be offered for many years.

Bowers' Educational Reforms in Higher Education

From my experiences at the EJ retreats, Elders Gatherings, AERA pre-seminars, and EJ conferences, my colleagues and I developed an integrated approach to educational reform. The article was published in 2009 in the currently obsolete *Ecojustice Education Review: Educating for the Commons* journal titled, *Beyond Dualism: Toward a Transdisciplinary Indigenous Environmental Studies Model of Environmental Education Curricula*. We developed an integrated approach to help capture the essence of the importance of infusing IK and EJE in teacher education. What follows is an excerpt from the article that was subsequently republished in *Contemporary Studies in Environmental and Indigenous Pedagogies: A Curricula of Stories and Place* (Kulnieks, Longboat, & Young, 2013a):

> Environmental education curricula in North America is primarily based upon the scientific model of inquiry. In an age of

unprecedented environmental degradation resulting in the loss of biodiversity, exponential population growth, sustainability questions, and as global climate change continues to soar to daunting heights, environmental education is failing to interpret the status quo that necessitates change. A paradigm shift, with a purpose to develop a model that bridges and seeks to integrate both academic disciplines and cultural knowledge systems into a more "integrative" process is needed to address environmental complexity. In this paper we outline the distinction between a scientific and Indigenous approach to environment and propose a transdisciplinary Indigenous environmental studies curriculum model enacted in an environmental studies program and in faculties of education through an ecojustice framework.

(Kulnieks et al., 2013a, p. 9)

Figure 9.1 illustrates a bridging of science and IK into an emerging field of Indigenous environmental studies.

Implications of this research beckon a return to the origins of environmental education including the natural and experiential approach in *corroboration* with scientific inquiry through an Indigenous environmental studies model enacted not only in emerging environmental studies programs but also in faculties of education. The knowledge that emerges from discussions with Indigenous Elders is a technique that investigates stories that are embedded in particular and specific places throughout North America. The anticipation is a focus of bridging ecologically sustainable knowledge and practices of both Indigenous and settler culture. Sharing knowledge about the local environments can provoke an informed dialogue about environmental issues that society as a whole was and continues to be facing. The significance of this work is that it provides opportunities for educators to become aware of an emergent field of environmental learning that includes an Indigenous understanding of the ecology of place and its relationship to human beings in that place.

(Kulnieks et al., 2013a, pp. 15–16)

As a transdisciplinary framework, it brings together IK & EJE into educational reform. The collection invites the reader into a conversation about the multiple dimensions of the relationship between EJE and Indigenous pedagogies.

In 2011 Bowers' published *Perspectives on the Ideas of Gregory Bateson, Ecological Intelligence, and Educational Reforms*. Bowers' publications helped to inform my ever-evolving understanding of his call for higher educational reform. For example, I worked to articulate the two primary

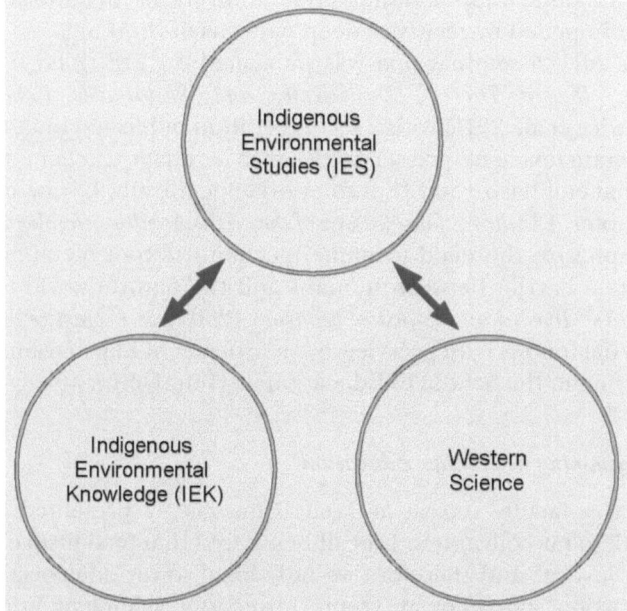

Figure 9.1 Relationships among Indigenous environmental knowledge, Indigenous environmental studies, and Western science

tasks of EJE—a cultural analysis of the ecological crisis and identifying diverse cultural educational methods to reveal how language continues to be a barrier in the relationship between humans and the natural world. Bowers' (2011) draws on the theories of Gregory Bateson to help articulate what lies at the very heart of EJ educational reform as he wrote:

> Bateson was attempting to resolve the problem of the relationship between the mind and the external world—a problem that is ignored by nearly all public school teachers and most university professors even though their task is to provide the conceptual frameworks that will guide how students think about the external world, as well as their internal world. (p. 25)

In order to express his understanding about the importance of learning about the (lack of) relationship between humans and the natural world, Bateson used the metaphors of the *map* and the *territory.* Bowers' stated: "The map refers to the cultural/metaphorical language/thought connections, while the 'territory' is the world of the natural and cultural systems we commonly refer to as the environment in which we live" (p. 25). Bower's taught his students to notice the problematic ways in which language perpetuates a taken for granted notion of progress that is linear in nature. He also taught us to notice the ways in which there are environmental limits and that our relationships with natural systems have been impeded by recursive deep patterns of thinking.

Also in 2011, a seminal text was published on EJE titled, *EcoJustice Education: Toward Diverse, Democratic, and Sustainable Communities* (Martusewicz et al., 2011) with a second edition published in 2015. I use this important text in my pre-service English language teacher education courses that emphasize an EJE framework. Specifically, I draw attention to the chapter *A Cultural foundations of the crisis: A cultural/ecological analysis* to emphasize the need to name mechanistic root metaphors that perpetuate a barrier between humans and the natural world together with Bowers' *Toward an eco-justice pedagogy* (2002) & *Ecojustice dictionary* (2004). What follows is an overview of an infusion of EJ programming in my department, the School of Education at Trent University.

EJE Programming in Teacher Education

In 2006 at a faculty retreat at Trent University, I presented Bowers' framework to my colleagues. I recall being told that "eco-justice" is part of "social justice" and therefore we only need to include social justice in our mission statement at Trent University's School of Education. However, I strongly believe that we need to name "eco-justice" in order that it may not be rendered invisible. I was successful in my articulation of the importance of including an "eco-social justice" framework in our mission statement. The inclusion of the word "eco" is at the heart of Bowers' framework as it is in the naming of the ways in which language carries forward past patterns of taken-for-granted thinking.

After the retreat, a group of eco-minded faculty began to discuss how we could infuse an EJE framework into our pre-service program. We embraced a paradigm that seeks to integrate academic disciplines with cultural knowledge systems (Beckford, Jacobs, Williams, & Nahdee, 2010) as such an approach is needed to address the complexity of environmental education and to inspire a new generation of teachers and their students. This was particularly pertinent as we were in a time of the UNESCO Decade of Sustainable Education and appeals for action at various levels in Canada had been made. At the national level, for example, the *Canadian Accord on Indigenous Education* (2010) called for teacher education to respond to the needs of Indigenous peoples, and at the provincial level, the *Ontario First Nation, Métis, and Inuit Education Policy Framework* (2007) and for environmental education, *Acting Today, Shaping Tomorrow: A Policy Framework for Environmental Education in Ontario Schools* (2009b), as well as resources for teaching about Aboriginal peoples *Aboriginal Perspectives: A Guide to the Teachers' Toolkit* (2009a) called for more IK to be infused in the curriculum. Our faculty began to explore contemporary environmental education frameworks so as to advance a more integrated curriculum informed through EJE and IK.

Our framework for (re)conceptualizing environmental education through an interdisciplinary program for pre-service teacher educators involves outdoor education, ecojustice pedagogy, food sustainability, and Indigenous ways of knowing that seeks to encourage, support, and foster environmental educational leadership. As our environmental education framework is interdisciplinary in nature, so, too, is our literature review. Our literature review informs our research objectives in terms of advancing a framework that involves learning by story, drawing upon nearby nature to explore biodiversity, exploring a cultural analysis of the environmental crisis, and ultimately helping teacher candidates to develop environmental leadership skills through a role of advocacy for the environment.

Our literature review consists of three broad areas. Specifically, and first, a survey of the literature covers key environmental texts that include classics such as Aldo Leopold's (1978) *Land Ethic* and Rachel Carson's *Silent Spring* (1962), as well as environmental sensitivity research (Palmer, 1998; Tanner, 1980) that reveals that one of the most important ways to help students develop a sense of advocacy for the environment is to provide them with rich encounters with the natural world during childhood. This sentiment is echoed in E. O. Wilson's (1992) classic book, *The Diversity of Life*, calling for conservation of the richness and diversity of life for future generations. Furthermore, and framed as a way to address "nature deficit disorder," Richard Louv (2008), in his acclaimed book *Last Child in the Woods*, outlines the importance of children having the opportunity to play in nature.

Second, our framework is also informed by literature that advances an understanding of storytelling and narrative inquiry as vehicles for environmental education (Cheney, 2002; Fawcett, 2002; Hart, 2002; Johnston, 2002; Moore, 2002), and an understanding of the importance of a connection between childhood development and the ecology of imagination (Cobb, 1959; Egan, 1986). As such, it is also informed by the work of theorists who understand the importance of learning by story and nature (Berry, 1991; Dooling & Jordan-Smith, 1989; Postman, 1989; Shepard, 1977; Sheridan, 1994) and those who examine the modern environmental movement and First Nations environmental wisdom (Suzuki, 1997; Suzuki & Knudtson, 1992).

Finally, the environmental education framework is informed by ecojustice educational research that highlights the importance of a cultural analysis of the environmental crisis (Bowers, 2002, 2006, 2011; Martusewicz et al., 2011, 2015; Longboat, Kulnieks, & Young, 2009, 2013), informed by indigenous environmental educational scholarship (Basso, 1996; Cajete, 1994; Chase, 1993; Laduke, 1999; McGregor, 2004) that advances the importance of infusing environmental education with IK in school curricula.

In addition to infusing EJE and IK in pre-service courses, in 2011, my colleagues and I co-developed an *Eco-mentorship Certificate Program* to support learners in their endeavor's to become environmental leaders. The program is cross-curricular as it brings together Indigenous, scientific, ecojustice, and outdoor education into an eco-mentorship curricula. The program itself involves collaboration between the School of Education, Camp Kawartha, and the Peterborough Community Gardens Initiative. The program supports pre-service teacher candidates engaged in environmentally focused activities for environmental leadership development, who are committed to problematizing and questioning the possibilities and limitations of the Ontario government's policy framework for environmental education *Acting Today, Shaping Tomorrow* (2009b). The program draws upon several eco-related themes.

For example, one of the themes is *Drawing on Nearby Nature*. As mentioned earlier, Environmental sensitivity research (Palmer, 1998; Tanner, 1980) has revealed that one of the most important ways to help students develop a sense of advocacy for the environment is to provide them with rich encounters with the natural world during childhood by exploring biodiversity. During the program, a variety of hands-on strategies and techniques for using nearby nature areas as a venue for environmental education are presented involving a cross section of environmental games, activities and resources relating to science, social studies, physical education and the arts. The intent is to inspire students to adopt an ethic of care and stewardship for their local environment.

A second theme is *Removing Barriers to Environmental Education*. As we know, educators are often full of good intentions. Yet the idea of teaching environmental education as well as all of the other subject areas is a bit overwhelming for many teachers. This theme addresses how to obtain administrative support for outdoor excursions, where to access funding and what local resources may be available to help deliver environmental education in the local community. During this part of the program, an EJE framework is introduced and there is a discussion about the ways in which language can hide and illuminate our relationships and understandings of the natural world. An overview of Bowers' framework is presented and remains integral to the program.

A third theme is *Environmental Education: Inspiring Hope*. By its very nature teaching environmental education can be both frightening and daunting to pre-service teachers. This theme addresses how we can teach students in ways that are both age appropriate and inspire hope for action. We need to, as environmental educator David Sobel (1996) suggests, find ways to go beyond "ecophobia" or the fear of facing an uncertain future and impending environmental disaster. It is here where we discuss our relationships to the natural world and any fears that we may have. As many Teacher Candidates are from urban areas, exploring the wilderness may be intimidating.

The final theme is *Indigenous and Environmental Education across the Curriculum*. With a jam-packed curriculum, it is not surprising that teachers often say: "I'd love to do environmental education but I just don't have the time." An integrated approach to environmental learning can effectively cover a number of expectations in a variety of subject areas. We want teachers to recognize that an Indigenous and environmental education perspective and related activities does not mean crowding out other subject learning. This theme addresses Indigenous teachings as well as an exploration of the benefits of setting up eco-activities (clubs, themed days, blogs, and competitions). Successful models of classroom practice that are experiential in nature and bring together meaningful learning in a variety of domains (especially integrating arts and science) are explored.

The four themes emerged from conversations exploring what current environmental education *is* and *looks like* in the field. Much of the contemporary view of the environment is shaped by a largely Eurocentric view of ecology (Battiste & Henderson, 2009). In response, environmental imperatives have suggested that environmental education is put back on track and returned to its origins, as reflected in David Orr's (1992) understanding that "all education is environmental education" (p. 90). The aim of the program is to consider how effectively an eco-mentorship education program prepares Teacher Candidates for environmental leadership and at the same time advances a framework of best-practice for integrated environmental learning that involves environmental leadership development. The consensus is that a paradigm shift, for the purpose of developing a framework that seeks to integrate academic disciplines and cultural knowledge systems into a more "integrative" process is needed to address the complexity of environmental education and to inspire a new generation of teachers and their students. The Eco-mentorship program provides 12 hours of curriculum for pre-service teachers.

In order to expand on the already existing *Eco-Mentorship Program*, in 2013, my colleagues and I developed a partnership between Ecology Park in Peterborough, Ontario and my department, the School of Education, at Trent University when we introduced the *Learning Garden Program*. Ecology Park is situated in Peterborough on a 5-acre parcel of sustainable landscape that is used as an outdoor teaching classroom (https://www.greenup.on.ca/ecology-park/). In our Bachelor of Education program, Teacher Candidates must complete a 75-hour alternative settings placement. The objectives of the placement include providing Teacher Candidates with the knowledge, motivation, and skills to facilitate the transmission of an environmental consciousness to their future students. Additionally, it assists Teacher Candidates in learning about eco-literacy, an EJE framework, environmental leadership, and food sustainability. Ten hours of workshops are provided to help to outline an ecojustice framework and engage in traditional teachings from local Elders. Teacher candidates also experience planning and garden-based activities at Ecology Park in our local outdoor classroom.

Pre-service teachers make connections to curriculum and teach a wide variety of K-12 classes that visit the outdoor education teaching park on full-day field trips. Pre-service teachers learn and teach about, for example, habitats and communities: Food chains, interdependence of plants and animals, trees, rocks and minerals; biodiversity, invasive species and interrelationships between/within species; habitats as areas that provide plants and animals with the necessities of life (e.g., food, water, air, space, soil and light); food chains as systems in which energy from the sun is transferred to producers (plants) and then to consumers (animals); balancing human needs and environmental stewardship; an ecosystem (e.g., a log, a pond, a forest) as a system of interactions between living organisms and their environment; permaculture ethics by working with what is in nature already; phenology in urban ecology by studying cyclical and seasonal natural phenomena; seasonal cycles and ecology of living systems of which we are an integral part of; and finally, biodiversity and resilience, earth care, fair share, food literacy, pollinators; and exploring growth cycles including composting.

Pre-service teachers engage in environmental education and make connections to curriculum through language arts, math, science and social studies, physical and health education, and the arts. Both the Eco-mentorship and Learning Garden programs continue to run at Trent University providing a unique opportunity for pre-service teachers to help foster greater environmental educational leadership as they are introduced to EJE & IK. What follows is an overview of our disseminated research in EJE.

Disseminating EJE Research

Throughout my EJE journey, in addition to disseminating our integrated framework (Kulnieks et al., 2009, 2013a), my research colleagues and I advanced an eco-hermeneutic approach to curriculum (Kulnieks et al., 2010). "An eco-hermeneutic curriculum includes moving beyond exclusively print-centered forms of learning in order to develop a deeper understanding of place" (p. 17). We subsequently explored Indigenizing environmental education curricula (Kulnieks et al., 2011), considered curriculum that fosters educational leadership (Kulnieks et al., 2013b, 2013c), reconsidered Canadian environmental curriculum studies (Kulnieks, Ng-A-Fook, Stanley, & Young, 2012), and explored educating teachers about IKs through an EJE framework (Kulnieks et al., 2013d),

Other EJE related publications include engaging in an arts-informed ecojustice pedagogy (Kulnieks & Young, 2014a); fostering a curriculum of ecological awareness through poetic inquiry (Kulnieks & Young, 2014b); developing an ecologically sustainable language arts curriculum via oral history education and poetic inquiry (Kulnieks et al., 2016); integrating indigenous curriculum through an ecojustice-arts-informed pedagogy and eco-hermeneutical methods (Kulnieks et al., 2016); exploring perspectives on environmental education through an eco-justice mentorship program (Young & Stanley, 2018); ruminating on the metaphorical nature of language (Stanley & Young, 2011; Young, 2018); developing ecological literacy through an integrated model as a primer for EJE curricula (Kulnieks et al., 2018); exploring ecological literacy in teacher identity (Kulnieks & Young, 2018); and developing a pedagogy for reconciliation and ecojustice-oriented education (Longboat et al., 2019). As I reminisce about my 15-year journey, I realize that the basis for all of the disseminated work is Bowers' (2002, 2006, 2011) deep understanding of the patterns of interdependence that all diversity depends upon, the need to recognize individuality is not autonomous, change is not a linear form of progress, and that nature is not a resource for commodification. While dissemination is an important part of the work in higher education, the curricular practices that are shared in our Teacher Education programs are as well and it is my commitment to EJE to help students to recognize the implications of our commodified culture and to recognize the deep cultural roots of ecological crisis. What follows is an overview of an EJE infused English language pre-service curriculum as I focus on Bowers' (2002, 2006, 2011) attention to root metaphors that reproduce hierarchies of deep patterns of thinking.

Teacher Education Reform—Identifying Diverse Cultural Methods for EJE Practices

In my close study of EJE with Bowers and his seminal written work, I had the opportunity to learn about how I can inspire curricular and pedagogical changes that involve drawing upon an ecojustice framework in teacher education. Specifically, in my English language classes, I implemented curricula that ask pre-service teachers to challenge cultural assumptions about their own relationships with the natural world. I wrote:

> I am ultimately proposing a reconceptualized curriculum that involves identifying practices that help humans get over an obsession with development, progress, and consumerism and consider how dominant cultural narratives perpetuate anti-ecological habits of mind. Dominant cultural narratives include taken-for-granted notions of human domination over the natural world through a commodification of "natural resources" into a celebrated story of "progress" based on a myth that consumption of these resources are endless and have no boundaries. These narratives are featured every day in media, conveying a message that bigger and more is better, new and improved products are superior to older ones, and

that progress is the inevitable trajectory of civilization while tradition is somehow backward.

(Young, 2009, p. 312)

Diverse cultural methods for EJE practices involve identifying language
barriers between humans and the natural world through an analysis of
our own ecological habits of mind (Young, 2008) through an EJ Action
Project that promotes an analysis of metaphorical language and our
relationship with the local cultural and ecological commons:

> As part of an eco-justice action project, students investigate the dif-
> ferences between past traditional ways of knowing and being with
> contemporary practices through an analysis of various cultural
> practices. Students also compare traditions of Indigenous peoples
> with non-Indigenous elders whether through oral storytelling, film,
> or storybooks. In Western society, each cultural practice (food,
> arts, etc.) is embedded in a larger dominant cultural narrative that
> privileges progress and consumerism over traditional experiential
> practices. In naming these cultural practices, students will learn to
> recognize cultural patterns (Bowers, 2006). Connections to domi-
> nant cultural narratives in media can then be explored and com-
> pared. (p. 316)

Analyzing cultural practices:

> Analyzing cultural practices. … involves comparing past and pres-
> ent ways of knowing and being. For example, when addressing the
> difference between approaches to traditional food preparation and
> today's fast food consumption, consider the following guiding ques-
> tions: How is food preparation and consumption different? Why
> is it different today? How do Indigenous peoples approach food
> preparation? Students can share their food stories. In addition, fur-
> ther integration of eco-justice vocabulary concerning food could
> address food growth, distribution, and waste (composting, garden-
> ing, planting, soil nutrients, fertilizers, harvest, preserving, and
> preparing food) through an investigation into local versus global
> approaches to food production. (pp. 316–317)

Comparing an oral vs print-based society:

> When addressing the difference between human relationships in an
> oral society and print-based society, consider the following guiding
> questions: What is the difference between an oral traditional soci-
> ety and a print-centric society? How have relations been changed
> and mediated via the computer and Internet? What methods of

communication did elders use and why? How are these different from contemporary modes of communication? Children can compare and classify these differences as a group with the teacher. (p. 317)

Comparing historical and contemporary geography:

> Other considerations include comparing historical local architecture and landscape with contemporary geography and mapping out the visual and aerial aspects and physical layout of community (past and present). This will naturally move into a consideration of what forms of enclosure are part of everyday life—water, airways, woodlots and pastures, privatized land, state ownership, patents, works of art, Internet, land roadways, and so on. Consider the following guiding questions: Where are greenhouses in the community? Where are farms located? How have land, water, airways, pastures, and so on been enclosed and privatized? (pp. 317–318)

Ultimately, we engage in activities such as using comparisons to identify root metaphors through analogies (Martusewicz et al., 2011, 2015). Examples include: "How is the body like a clock or computers? (revealing a mechanistic root metaphor). How is it different? How is the body like a plant (repositioning the body through an ecological root metaphor?" (Adapted from Bowers teachings) (Young, 2009, pp. 313–314).

> We develop a repertoire of guiding questions and eco-justice vocabulary. For example: How is a stream like a drain? How is it different? This is an important comparison of two similar things that essentially act in the same manner (both move water toward a larger body of water). How do we treat these same things differently? Which word, *drain* or *stream* is derived from a mechanistic root metaphor? Why does it matter? Would you put different things in a stream versus a drain? These are examples of how language mediates relationships. (p. 314)

Another final example was mentioned earlier that involves environmental autobiographical writing practices to address our relationships with the natural world. This is extended into EJE and storytelling by exploring dominant cultural narratives. Ultimately, in addition to outlining ecojustice curricula and practices in teacher education programming, I have shared the ways in which Bowers' influence on my work has been extensive over the last 16 years. Bowers' mentorship has helped me to re-envision curriculum studies through a linguistic and cultural analysis of the social and ecological crises that involves ecological literacy and eco-hermeneutics. I am forever grateful for his mentorship.

References

Association of Canadian Deans of Education. (2010). *Accord on indigenous education*. Montreal: Authors.

Basso, K. (1996). Wisdom sits in places: Landscape and language among the Western apache. *Albuquerque*. NM: University of New Mexico Press.

Battiste, M., & Henderson, J. (2009). Naturalizing indigenous knowledge in Eurocentric education. *Canadian Journal of Native Education, 32*(1), 5–18.

Beckford, C. L., Jacobs, C., Williams, N., & Nahdee, R. (2010). Aboriginal environmental wisdom, stewardship, and sustainability: Lessons from the Walpole Island first nations, Ontario, Canada. *The Journal of Environmental Education, 41*(4), 239–248.

Berry, W. (1991). Out of your car, off your horse. *Atlantic Monthly*, February, 61–63.

Berry, W. (1995). *Another turn of the crank*. Washington, DC: Counterpoints.

Borgmann, A. (1992). *Crossing the postmodern divide*. Chicago, IL: University of Chicago Press.

Bowers, C. A. (2001). *Educating for eco-justice and Community*. Athens, GA: The University of Georgia Press.

Bowers, C. A. (2002). Toward an eco-justice pedagogy. *Environmental Education Research, 8*(1), 21–34.

Bowers, C. A. (2004). *Ecojustice Dictionary* [cited September 1st 2005]. Available from: http://www.cabowers.net/dicterm/CAdict003.php

Bowers, C. A. (2006). *Revitalizing the commons: Cultural and educational sites of resistance and affirmation*. Lanham, MD: Lexington Books.

Bowers, C. A. (2011). *Perspectives on the ideas of Gregory Bateson, ecological intelligence, and educational reforms*. Eugene, OR: Eco-Justice Press.

Cajete, G. (1994). *Look to the mountain: An ecology of indigenous education*. Skyland, NC: Kivaki Press.

Carson, R. (1962). *Silent spring*. Boston, MA: Houghton Mifflin Company.

Chase, A. (1993). Traditional ecological knowledge: Wisdom for sustainable development. *The Australian Journal of Anthropology, 4*(3), 245–247.

Cheney, J. (2002). The moral epistemology of First Nations stories. *Canadian Journal of Environmental Education, 7*(2), 88–100.

Cobb, E. (1977). *The ecology of imagination in childhood*. New York, NY: Columbia University Press.

Dooling, D. M., & Jordan-Smith, P. (Eds.). (1989). *I become part of it. Sacred dimension in Native American life*. San Francisco, CA: HarperCollins.

Ecology Park. https://www.greenup.on.ca/ecology-park/.

Egan, K. (1986). *Literacy, society, and schooling: A reader*. Cambridge, UK: Cambridge University Press.

Fawcett, L. (2002). Children's wild animal stories. *Canadian Journal of Environmental Education, 7*(2), 125–139.

Hart, P. (2002). Narrative knowing, and emerging methodologies in environmental education research. *Canadian Journal of Environmental Education, 7*(2), 140–165.

Internet Archive: https://web.archive.org/web/20100919120531/http://ecojusticeeducation.org/index.php?option=com_content&task=view&id=67&Itemid=44.

ISEC. (1993). *Ancient Futures: Learning from Ladakh.* Video Project. www.isec.org/uk/ladakh Accessed September 12, 2005.

Johnston, R. (2002). Wild Berwyn or Coy Nature Reserve. *Canadian Journal of Environmental Education, 7*(2), 166–178.

Kulnieks, A., Longboat, D., & Young, K. (2010). Re-indigenizing learning: An eco- hermeneutic approach to curriculum. *AlterNative: An International Journal of Indigenous Peoples, 6*(1), 15–24.

Kulnieks, A., Longboat, D., & Young, K. (2011). Indigenizing curriculum: The transformation of environmental education. In D. Stanley & K. Young (Eds.), *Contemporary studies in Canadian curriculum: Principles, portraits & practices* (pp. 351–374). Edmonton, AB: Brush Education.

Kulnieks, A., Longboat, D., & Young, K. (2013a) *Contemporary studies in environmental and indigenous pedagogies: A curricula of stories and place.* Brill: Sense.

Kulnieks, A., Longboat, D., & Young, K. (2013b). Eco-literacy development through a framework for indigenous and environmental educational leadership. *Canadian Journal of Environmental Education, 18,* 112–126.

Kulnieks, A., Young, K., & Longboat, D. (2013c). Indigenizing environmental education: Conceptualizing curriculum that fosters educational leadership. *First Nations Perspectives – The Journal of the Manitoba First Nations Education Resource Centre, 5*(1), 65–81.

Kulnieks, A., Longboat, D., & Young, K. (2013d). Engaging literacies through ecologically minded curriculum: Educating teachers about indigenous knowledges through an ecojustice education framework. *in education, 19*(2), 138–152.

Kulnieks, A., Longboat, D., & Young, K. (2016). Engaging eco-hermeneutical methods: Integrating indigenous curriculum through an eco-justice-arts-informed pedagogy. *AlterNative: An International Journal of Indigenous Peoples, 12*(1), 43–56.

Kulnieks, A., Longboat, D., & Young, K. (2018). Developing eco-literacy through an integrated model: A primer for eco-justice education curricula. In *Eco-justice: Essays on theory and practice in 2017* (pp. 11–22). Eugene, OR: Eco-Justice Press.

Kulnieks, A., & Young, K. (2018). Exploring ecological literacy in teacher identity: Reflexive inquiry through a learning garden curricula. In E. Lyle (Ed.), *The negotiated self: Employing reflexive inquiry to explore teacher identity* (pp. 76–85). Brill: Sense.

Kulnieks, A., Longboat, D., Sheridan, J., & Young, K. (2016). Oral history education through poetic inquiry: Developing ecologically sustainable language arts curriculum. Our schools/our selves. *Canadian Centre for Policy Alternatives Quarterly Journal on Education, 25*(2), 127–134.

Kulnieks, A., Ng-A-Fook, N., Stanley, S., & Young, K. (2012). Reconsidering Canadian environmental curriculum studies. In N. Ng-A-Fook & J. Rottmann (Eds.), *Reconsidering Canadian curriculum studies* (pp. 107–136). New York, NY: Palgrave MacMillan.

Kulnieks, A., & Young, K. (2014a). Ekphrastic poetics: Fostering a curriculum of ecological awareness through poetic inquiry. *in education, 20*(2), 79–89.

Kulnieks, A., & Young, K. (2014b). Literacies, leadership, and inclusive education: Socially justice arts-informed eco-justice pedagogy. *LEARNing Landscapes, 7*(2), 183–194.

Laduke, W. (1999). *All our relations: Native struggles for land and life – selection.* Boston, MA: South End Press.

Leopold, A. (1978). *A sand county almanac, and sketches here and there.* New York, NY: Oxford University Press.

Longboat, D., Kulnieks, A., & Young, K. (2009). Beyond dualism: Toward a transdisciplinary indigenous environmental studies model of environmental education curricula. *The EcoJustice Review: Educating for the Commons, 1*(1), 1–18.

Longboat, D., Kulnieks, A., & Young, K. (2013). Beyond dualism: Toward a transdisciplinary indigenous environmental studies model of environmental education curricula. In A. Kulnieks, D. Longboat, & K. Young (Eds.), *Contemporary studies in environmental and indigenous pedagogies: A curricula of stories and place* (pp. 9–18). Rotterdam, Netherlands: Sense/Brill. ISBN: Previously published in *The EcoJustice Review* (EJR, 2009) Internet Archive: https://web. archive.org/web/20100919120531/http://ecojusticeeducation.org/index. php?option=com_content&task=view&id=67&Itemid=44

Longboat, D., Kulnieks, A., & Young, K. (2019). Developing curriculum through engaging oral histories: A pedagogy for reconciliation and eco-justice-oriented education. In K. Llewellyn & N. Ng-A-Fook (Eds.). *Storying historical consciousness in times of reconciliation: Oral history, public education, and cultures of redress* (pp. 183–196). New York, NY: Routledge.

Louv, R. (2008). *Last child in the woods: Saving our children from nature-deficit disorder.* Chapel Hill, NC: Algonquin Books of Chapel Hill.

Martusewicz, R. A., Edmundson, J., & Lupinacci, J. (2011). *EcoJustice education: Toward diverse, democratic, and sustainable communities.* New York, NY: Routledge.

Martusewicz, R. A., Edmundson, J., & Lupinacci, J. (2015). *EcoJustice education: Toward diverse, democratic, and sustainable communities* (2nd ed.). New York, NY: Routledge.

McGregor, D. (2004). Coming full circle: Indigenous knowledge, environment and our future. *American Indian Quarterly, 28*(3–4), 385–410.

Moore, J. (2002). Lessons from environmental education. *Canadian Journal of Environmental Education, 7*(2), 179–192.

Orr, D. (1992). *Ecological literacy: Education and the transition to a postmodern world.* Albany, NY: State University of New York Press.

Palmer, J. A. (1998). *Environmental education in the 21st century: Theory, practice, progress, and promise.* New York, NY: Routledge.

Pinar, W. ed. (2004). *Contemporary curriculum discourses: Twenty years of Jct.* New York, NY: Peter Lang.

Postman, N. (1989). Learning by story. *Atlantic Monthly,* December, 119–124.

Shepard, P. (1977). Place in American culture. *North American Review,* Fall, 22–32.

Sheridan, J. (1994). Alienation and integration: Environmental education in Turtle Island. Unpublished doctoral dissertation, University of Alberta, Edmonton, AB.

Shiva, V. (1993). *Monocultures of the mind: Perspectives on biodiversity and biotechnology.* Chicago, IL: Third World Network.

Sobel, D. (1996). *Beyond ecophobia: Reclaiming the heart in nature education.* Great Barrington, MA: Orion Society.

Stanley, D., & Young, K. (2011). Conceptualizing the complexities of curriculum: Developing a lexicon for ecojustice and the transdisciplinarity of bodies. *Journal of Curriculum Theorizing, 27*(1), 36–47.

Suzuki, D. (1997). *The sacred balance: A vision of life on earth.* Vancouver, BC: Greystone Books.

Suzuki, D., & Knudtson, P. (1992). *Wisdom of the elders.* Toronto, ON: Stoddart Publishing.

Tanner, T. (1980). Significant life experiences: A new research area in environmental education. *Journal of Environmental Education, 11*(4), 20–24.

Wilson, E. O. (1992). *The diversity of life.* Cambridge, MA: Harvard University Press.

Young, K. (2005). Developing ecological literacy as a habit of mind in teacher education. *The EcoJustice Review: Educating for the Commons, 1*(1), 1–7. Internet Archive: https://web.archive.org/web/20101031082234/http://ecojusticeeducation. org/index.php?option=com_content&task=view&id=35&Itemid=46

Young, K. (2008). Ecological habits of mind and the literary imagination. *Educational Insights,* 12(1), 1–9.

Young, K. (2009). Reconceptualizing early childhood literacy curriculum: An ecojustice approach. In L. Iannacci, & P. Whitty (Eds.), *Early childhood curricula: reconceptualist perspectives* (pp. 299–325). Calgary, AB: Detselig/Brush Education.

Young, K., & Stanley, D. (2018). Integration, inquiry, and interpretation: A learning garden alternative placement & eco-mentorship program for pre-service teachers. In G. Reis & J. Scott (Eds.), *International perspectives on the theory and practice of environmental education: A reader* (pp. 47–56). New York, NY: Springer.

Young, K. (2018). The character of contemporary curriculum studies in Canada: A rumination on the metaphorical nature of language. In E. Hasebe-Ludt & C. Leggo (Eds.), *Provoking curriculum studies: A métissage of inspiration/imagination/ interconnection* (pp. 78–84). Toronto, ON: Canadian Scholars' Press.

www.bowers.net/ecojusticedictionary.

10 Coyote and Raven Encounter Chet Bowers in Conceptual Time-Space

Ecojustice Pedagogies of the Land

Peter Cole and Pat O'Riley

Based on our own cultural knowing as Indigenous scholars and over two decades of research with Indigenous communities in Canada and more recently in the High Amazon of Peru, we join this conversation on the work of Chet Bowers in the shape of an oral [speaking on the page] narrative that journeys across and between the Global South and Global North. We look at some of Bowers' key concepts that resonate with the voices and lived experiences of Indigenous Peoples, who, like Bowers (2013), want to see "a broader understanding that represents the environment as encompassing the global cultural forces that are contributing to a sustainable future, as well as those that are accelerating the destruction of the natural systems leading to the collapse and disappearance of cultures" (p. 225).

Bowers acknowledges that there is much to learn from Indigenous knowledges. Our own Indigenous co-researchers want to have their voices heard so that their self-determining cultural regeneration efforts are supported, their lands protected from expropriation and exploitative resource extraction, and their ecological knowledges valued in the academy.

With our Indigenous co-journeyers, we deliberate on several of Bowers' "root metaphors" (2012) or "metacognitive schemata" (Bowers, 2005) with the hope of generating an inclusive and engaged cross-cultural conversation on ecojustice education beyond the "progress narrative" of modernity. For example, we ask: What might the dimensionalities, potentialities and collectivities of "cultural commons" become if inhabited by not only humans, but also non-humans and more-than-humans? How might this reshape Bowers' concept of "enclosure"—a concept that refers primarily to human cultures that are deeply situated in the Western industrial/consumer-dependent mindsets?

Furthermore, setting up our narrative journey we ask: How might Bowers' notion of "ecological intelligence" be expanded beyond awareness of impacts of local activities in the ethnosphere and biosphere if human, non-human, and more-than-human intelligences and agencies were integral parts of the "cultural commons?" And, especially if they were not Western industrial/consumer-dependent?

Bowers' (2013) writing about the potential of environmental education undermining Indigenous education, was concerned with "linguistic colonization" created through Enlightenment English-language-based and print-based communication. Considering this, we ask: In a culturally diverse commons, how might the narrativity, orality and performativity of Indigenous Peoples animate and enhance non-anthropocentric communications?

Bowers was also troubled by educators' uncritical promotion of digital literacies, another form of communication that propels students into virtual(ly), culturally and linguistically homogenous and essentialized worlds, disconnecting them from the messy, embodied, material, social, cultural, ecological "real world" issues and sensitivities. Considering his critique together with our concerns, we ask: How might Indigenous land-based knowledges and traditional practices address some of Bowers' linguistic and digital colonization concerns in education while also re-awakening the human spirit and attentiveness to the sentient more-than-human entities, intelligences and interdependencies? And, in what ways might Indigenous ecoliteracies, ecotechnologies, and ecopedagogies contribute to re-generating more complex, compassionate, and culturally inclusive possibilities for living together on a shared planet?

Bowers (2005) defined an important metaphor in his work "ecology." He wrote:

> The metaphor *ecology* has been taken over by the sciences, which may lead many readers to think that my reference to a cultural ecology is an inappropriate use of the term. However, if we go back to the original meaning of the ancient Greek word "oikos," before Ernst Haeckel in 1866 transformed it into "oecologie" and used it to represent the study of the natural environment, it referred to the operations and management of the family household. By recovering the ancient Greek meaning of the word, it is possible to avoid the limitations of viewing intelligence as an attribute of the autonomous individual. The metaphor of "household" can be extended to the larger patterns of interaction and interdependence that we find in community, culture, and the way in which cultures are embedded in natural systems. (p. 158)

Following such a foundational understanding of interdependence, Bowers argues that students need to understand how the root metaphors and processes of Western scientific thinking of modernity frame and act as foundations for their current ways of thinking and acting. This awareness would encourage attentiveness to their culture-language-thought connections, radically changing how they think about the content of their received schooling and their roles and responsibilities in an inter-connected and unequal world. Bowers (2002), wrote:

> The constitutive role of root metaphors in framing thought can be summarized by paraphrasing Nietzsche and Heidegger in the fol-lowing way: language thinks us as we think within the conceptual categories that the language of our cultural group makes available. As thought is inherently metaphorical, there is always the possibility of identifying more adequate analogies, and even of recognizing aspects of cultural/personal experience that previously held root metaphors cannot account for. (p. 23)

Bowers was also quite concerned with intergenerational learning and critical pedagogy. Working toward an ecojustice pedagogy, Bowers (2016) wrote:

> What needs to be avoided is exposure to a curriculum that deni-grates their heritage of intergenerational knowledge—which may include elder knowledge, patterns of mutual aid and solidarity that link together extended families and community networks, ceremo-nies, narratives, and other traditions essential to their self-identity and moral codes. (p. 28)

In his criticism of critical pedagogy in relation to ecological literacy, Bowers (2002) wrote:

> In order to recognize the profound difference in the root meta-phors that underlie an eco-justice pedagogy, it is first necessary to identify the root metaphors that are taken-for-granted by critical pedagogy theorists. Their root metaphors ... were the basis of the Industrial Revolution that is being continued today under the new metaphor of "globalization," which has replaced the older and highly criticized metaphor of "colonization.... Contrary to their claims, the practice of a critical pedagogy does not lead to individual eman-cipation and social justice; rather it reinforces a subjectively centered individualism required by the consumer, technologically-dependent society. While they are highly critical of the capitalist foundations of society, the root metaphors that underlie their prescriptions for change create a double bind they fail to recognize." (p. 26)

Bowers, continues:

> [T]he grammar of critical discourse advantages groups ... who have the power to dictate the rules governing what constitutes legitimate speech.... This dismisses speech of traditional societies that include

mythopoetic narratives that may be the foundation of a cultural group's moral codes, intergenerational knowledge that carries forward an understanding of the limits and possibilities of the bioregion, wisdom of elders, and forms of knowledge that come from direct experience of negotiating relationships in everyday life.... The key point here is that the rules governing critical inquiry are elitist in that they do not allow for the voices of cultural groups that do not assume that change...is always progressive in nature. (p. 28)

We invite you to join our trickstering co-journeyers Coyote and Raven and ourselves, in this word/body/mind/spirit exchange with Chet Bowers into currents of inter- and intra-being/becoming as we converse about how we might live together on a shared planet.

Coyote and Raven have been watching the Doomsday Clock (code name "trinity") tick and tock whirr and zing since it was first curated in the *jornada del muerto* desert in New Mexico by the makers of the atomic bomb—the Alamogordo Manhattan project—the ticking started July 16, 1945, Monday morning before dawn though in reality it has been thousands of years of warfare in the making.

voiceover "Peace, break thee off; look, where it comes again!" (Shakespeare, 1603/1623).

coyote says strange voices odd apparitions we might have been zapped by this armageddonic space-timepiece that runs on atomic radionucleotides   there's no r-value between it and us   it depends which doomsday clock model you refer to   this one uranium-235 is mechanistic and has a half-life of roughly 700 million years which is fairly stable as radio-isotopes go depending on the how and what of human in(ter)vention

raven says I don't trust any product whose side effects include volunteer transmutative auto-bioluminescence and spontaneous fission I wouldn't want one in my parlour   but how do you feel about it conceptually

conceptual space is not a safe place for many ideas   *says coyote* there's always leakage into the living world the words are attached to   the viral dimension of language—anyway conceptual space is ground zero for many infective ideas founded on ratio-causal-logical templates

raven says I think our old friend chet bowers would agree that words can be suspicious things to put your trust in—words about words and words about silence   how does one vaccinate oneself against the virulence of language   how does one activate immunity

the ghost of chet bowers slowly walks the battlements separating corporate greed political corruption and baseline stupidity from the inspirited beings of land sea and air

"Murder most foul, as in the best it is; But this most foul, strange and unnatural" (Shakespeare, 1603/1623).

coyote says he's talking about premeditated biocide insecticide herbicide everything-icide that arises from the mainstream western religious-academic-political belief that 'man' has dominion over all of creation   nobody asked me says an octopus a jellyfish a cedar tree and an amoeba in unison

voiceover "Words words words" (Shakespeare, 1603/1623).

words!   *raven says* I've always put my trust in random sounds and unscheduled movements rather than thought and movement based on rational or logical planning

coyote asks what were you thinking of using to think with raven?   words are the usual furniture of conversation   do you not consense with other corvids in meaningful spontaneous sound events—not necessarily attributing your conversation to morpheme phoneme or logomorph?

no   sometimes and perhaps   *says raven*   not representation nor symbolism they turn everything into cliché

asks coyote how does one participate consensually in communicative intercourse without resorting to cliché generation?   each metaphor joins foregoing ones connecting time *no more* with time *not yet* even though time is at once nonexistent and eternal

the ghost moves about the battlements as if distracted

voiceover "What art thou that usurp'st this time of night ... by heaven I charge thee, speak!"   (Shakespeare, 1603/1623).

says coyote see how the ghostly figure returns much agitated [its visor up muttering to itself]   "nay, answer me: stand, and unfold yourself" (Shakespeare, 1603/1623).

chet says "The ecological crisis is not really being understood in terms of how immanent it is, the acidification of the world's oceans, the melting of glaciers, the droughts, the extreme weather conditions. These are all bearing down on us" (Bowers, 2015a).

yes   *says raven*   flying is becoming risky   even the best thermal and jet stream is worrying

coyote asks   raven my flighty friend   how does one address a ghost?   deferentially   whether or not the ghost has been anointed crowned knighted damed or canonized?   what voice person and mood does one use?

ah!   *says raven*   passive voice   third person singular (pause) subjunctive mood   I want to ask mr. bowers how the ecological crisis is related to language

says coyote   but one doesn't interrogate a ghost   remember one employs the subjunctive and only if sorely pressed   the imperative   be sure that your words bow or curtsy   as occasion demands

voiceover   "Thou art a scholar; speak to it" (Shakespeare, 1603/1623).

the ghost of Chet speaks slowly   halting ly   "The basic question is to what extent do the words that encode the earlier assumptions and ways of   thinking continue to influence our current way of thinking and so to get at that one needs to understand the metaphorical nature of language" (Bowers, 2015d).

coyote asks   how does one escape this metaphorical language?

the ghost gesticulates and speaks haltingly   "[W]e're born into those language communities and they become the basis of our thinking...We need to go back a step and look at how reality is really experienced in the world by all forms of life and that is that our reality is emergent, relational and codependent" (Bowers, 2015a).

says raven   it's a tall order to imagine how any other being experiences the world *the ghost continues*   "What are the effects of the Industrial Revolution on the environment, and how is the digital revolution changing the thought patterns of other people in the world? How is it interrupting their intergenerational communication?" (Bowers, 2015a).

raven asks   what kind of thinking can one use to extricate oneself from consensual realities that stifle us?

the ghost responds   "Our intelligence is being affected by the intelligence encoded in the selection of analogues that frame the meaning of words, abstract fixed representations of reality" (Bowers, 2015a).

raven waves a wand around   science tells us it can take the world apart and put it back together better than the original using only one way of thinking   one way of validating—the culture and knowledge of empire   science claims reality is reducible only to western theories hypotheses rules of evidence proof and truth regimes

says coyote   such an uncompromising attitude and cultural hubris would require considerable attitudinal shift to even recognize other cultural values and ways of knowing let alone consider them worthwhile saving

chet continues his soliloquy surrounded by swirling fog   "There is nothing that is permanent in the world. It's all undergoing change at various rates and through various languaging processes" (Bowers, 2015a).

coyote asks   if the same languaging is being used to justify validate and foreground the thinking that has caused climate change how can anything change?

says raven   I've long been partial to not thinking   it keeps you from falling into a perpetual state of unnecessary planning and   organizing   don't get me wrong   ravens engage in thinking but we more often rely on experiential   cultural and intuitive cues

says coyote   there's no use worrying about the inevitable   just make sure your mortgage and life insurance are up to date   and somebody turned the stove off and fed the cat

the ghost of chet bowers paces with a troubled brow   "Now everything has to be monetized and contribute to market" (Bowers, 2015a).

that's not comforting   *says raven*   who is holding a geiger counter near the doomsday clock   and a stethoscope against its base.   someone has over-wound this clock   it's moved closer to midnight with a hey ho! and a hey nonny!

coyote's ears prick up   don't worry about midnight being doomsday   it's a random metaphor

the ghost moves in the mist   pauses   "How is it that a root metaphor that is an interpretive framework that affects how we look at many aspects of our everyday life prevents us from looking at what is problematic? What is not included and can't be understood in terms of the paradigm of   endless innovation and technological progress?" (Bowers, 2015a).

coyote says digital technology is even affecting what words children can access in their dictionaries   technology words such as 'broadband' and 'voicemail' replacing nature words such as 'bluebell' and 'wren'   and even more worrisome raven   the word 'raven' from the *Oxford Junior Dictionary*   virtuality replacing breath and spirit   cyberworlds are the new cultural commons

the ghost is reading The Lost Words (Macfarlane & Morris, 2018)   "[O]f the dominant aspects of our taken for granted world is the way in which we rely upon the language of earlier generations who were unaware of environmental limits, who were unaware of other cultural ways of knowing" (Bowers, 2015a).

raven   *says coyote*   we know that an unlanguageable world lies beyond rhizomatic mycellial mycorrhizal metaphors and layered   stratified   contiguous and saltatory networks

so   *says raven*   when representation replaces or obscures action-on-the-ground   when consensus overshadows the process-of-a-life-being-lived a web becomes the fuzzy-logic worked-strands that invisibilize the being-in-the-world that exists in multiple modes of focus ongoingly through a process of oscillation and flow   at some point a lived-life needs to be on level with theory and scientific guesswork   it is here where words become spells   charms or curses

the ghost stares into space   "We need to look at our patterns. It's not a matter of going into theory or other abstractions, it's a matter of looking at our words, the analogs that frame the meaning of those words" (Bowers, 2015a).

asks raven   are you afraid of all time pieces?   do they all signal doom?

says coyote   not afraid   just uneasy around time-space pieces

raven asks   how about the town clock in München's townsquare   the *Rathouse Glockenspiel* in *Marienplatz*?   I understand it symbolizes the plague of 1517 through the *Schäfflertanz*?

coyote says   every clock is potentially a doomsday clock   the *Marienplatz* one is a memorial to the dancing barrelmakers who with the *Schäfflertanz* symbolize the end of the plague of 1517

yes   *chimes in raven*   a celebratory chronometer is more to my liking   ding dong happy things are happening

the ghost ambles along in its personal fog   "What it is that you can bring as a participant in the cultural commons that will become increasingly important in the years ahead? People who have nothing to give to a community   should be understood as impoverished even though they may be very wealthy in the old meaning of that word" (Bowers, 2015a).

voiceover   "a mark a yen a buck or a pound ... is all that makes the world go 'round ... money money money money   money money money money" (Cabaret Lyrics, n.d.)

the ghost speaks again   "The world that we live in as emergent, relational and codependent (pause)   we don't live in a world of fixed entities of permanence ... and it's because of the history of our language that we take for granted many of the misconceptions encoded in that language" (Bowers 2015b).

says raven   I'm not sure exactly what the doomsday clock is signifying by midnight something will end   it's good to have reminders and warnings but we need to actually have ideas for moving forward with a better attitude and ecological and ethical footprint

a chime   then the ghost speaks   "Whether the ecology of language with its history is adequate for understanding the ecology of the world we live in with its ecological crisis bearing down upon us" (Bowers 2015b).

though in a sense   *says coyote*   language can be specific   it is also always contingent because language rarely refers to itself or to happenings outside of it   we are to assume that everything is language   and language is a kind of truth-bearing regime

raven counters   we can live our lives as if they were made up of doom-moments   or we can do something to make things better

you sound like a self-help workshop facilitator   *says coyote*   I mean that in the nicest way

the ghost goes on   "I talked about image metaphors and how they encode the thought processes of earlier times when they settled upon certain analogues that framed the meanings of these words" (Bowers 2015a).

so   *says raven*   I heard this eco-academic in the coffee shop talking about language and environment   and the point of no return

coyote raises her eyebrows   doom scenarios!   there's a lot of lead content in words these days

the ghost takes a drink of water   "The book of Genesis has two root metaphors that are very pronounced   and that is it's a human centred and that it's a patriarchal world" (Bowers 2015b).

says coyote   the eurolanguages and cultures are patriarchalized to the gills   (holding up two quote digits on each paw and pointing to ghost) "Hindsight enables us to recognize how the root metaphor of patriarchy dictated how attributes of men and women were to be understood, and thus what behaviors were moral. The root metaphors of anthropocentrism, individualism, mechanism, and so forth, also reproduce earlier culturally specific ways of understanding relationships, the attributes of the participants in the relationships, and thus what constitutes the moral norms governing the relationships" (Bowers, 2005, p. 178).

interesting   *says raven*   but roots are not the only connectors and transmitters between the life of the earth   water and sky   there are myriad connections between all living spirited beings

says the ghost   "More recent root metaphors include the notion of individualism, progress, economism. That is what we understand   all relationships in terms of their economic potential or value. There's also the root metaphor of evolution. Ecology itself is a root metaphor" (Bowers, 2015c).

coyote rejoinders   that's a lot of talk about roots by people who are not so apt to get their hands or claws or beaks into the earth

the ghost adds   "Perhaps the most important root metaphor that I overlooked is the root   metaphor of mechanism. It starts with Johann Kepler who said my aim is to show that the celestial machine is to be likened not to a divine organism but to a clockwork. So Thomas Hobbes who gives us the foundations of the modern political system says   for what is the heart   but a spring   and the nerves   but so many strings   and the joints   but so many wheels giving motion to the whole body" (Bowers 2015b).

raven says   human beings have always been crazy about tools   it started with sticks and stones

the ghost speaks   "Then we have William Harvey telling us the heart is a pump so the medical world becomes oriented toward looking at the   body as a system of parts of organisms" (Bowers 2015b).

says coyote   you are a well-informed ghost   you must have a good research library

the ghost continues   "Marvin Minsky one of the early artificial intelligence gurus says conscious thought uses signals, signs to steer engines in our mind.

raven adds   E. O. Wilson says "the mind is a machine a process that needs to be re-engineered... He says another quality that successful genes will have is a   tendency to postpone the death of their survival machines   at least until after reproduction" (Bowers 2015b).   dna-based genes seem to be bent on leaving a legacy copy of themselves   the 'me too' factor   I want to live forever

the ghost speaks again   "Wilson says the machine the biologists have opened up is a creation of riveting beauty. So the question is in what ways has the root metaphor of mechanism influenced agriculture, education, medicine, most aspects of daily life where we look at things in terms of machine-like qualities that can be measured, that can be re-engineered, that can be assessed in terms of increased efficiency" (Bowers 2015b).

voiceover   "the problem with the root metaphor has to do with what it excludes because it lacks the vocabulary for naming these alternative realities   within the root metaphor of mechanism for example   there is no reference to what is sacred   there is no reference to the history of human experience   the unique qualities of human beings" (Bowers 2015b).

says coyote   maybe these metaphors worked as filler at one level of human development in one instance and became embedded at every level and nobody did an audit   or inventory *the ghost pauses says pointedly*   "The personal pronoun 'I' reflects the root metaphor of individualism that doesn't take into account the way in which thought patterns are carried forward from the past, not only of the individual but everybody else who then shares that notion. The notion of private property is connected to the notion of individualism that in turn is reinforced by the root metaphor of progress" (Bowers 2015c).

university president to board of governors and auditorium filled with donors how much harm can a root metaphor really do?   (silence)

coyote says   the resource extraction industry wealth comes from violence   intimidation bribery and theft of indigenous lands globally yet their names go up on university buildings in every city   what kind of awareness or sense of justice does that represent?   government and industry and academia with their hands in one another's pockets   statues are being toppled

says raven   the 'donors' supply 2% of the funding and the public supplies 98% but guess whose name goes on the building or grant or chair   and academic administrators speak in reverential tones of their 'generosity' out of willful ignorance   apathy   or collusion

says coyote   I said to chet over coffee   as trickster beings we have a limited number of yelps and yips and meta-language we use to share our complex inner world our physical   emotional mental   spiritual selves   and histories but they seem to work   for us

raven asks   what did he say?

coyote replies   he didn't   he thought I was just a coyote making coyote sounds

raven asks   what about these baseline philosophers whose ideas have become foundational in western and westernized societies   like aristotle plato socrates

chet smiles   "What they gave us were abstract accounts of a reality that was not part of anybody's cultural experience…a tradition of thinking in the abstract that was ethnocentric and did not take into account environmental issues of their day" (Bowers 2015c).

coyote chimes in   as much as we use language to help us think we let it lead us to act out clichéd lives   we experience our thoughts   not our lives   we have become slaves   simulacra   of words without thoughts   thoughtless words   all hail!   or snow

the ghost speaks in alarm   "My words fly up, my thoughts remain below. Words without thoughts never to heaven go" (Shakespeare, 1603/1623).

coyote howls   our words always betray us as we do battle alongside them   for/yet we are dressed in their grammar syntax   sonority etymology   rhythm tritenesses

says raven   even as we play with words it is they who play with us   composing us   language writes us   speaks us   we are   become   the written

chet's ghost speaks   "Even the way in which print-based consciousness shows up in our tendency to think and communicate with each other in highly abstracted language where there is no sense of context, no sense of the relational world that we are in" (Bowers 2015c).

says raven   we are one another's prisoner   sculpting one another and the land with words   destroying it with vision that is only myopia blindness   looking but not seeing

coyote says   raven you are a sad poet   what is it that is not being said   or is being ignored?

chet intervenes   "The silences that surround the notion of progress ... supporting vocabularies about change, innovation, new ideas and so on" (Bowers 2015c).

yes   *says coyote*   there are many silences and silencings

chet's ghost walks   pondering   "One of the silences is that traditions are important particularly now that the computer revolution is overturning so many traditions such as privacy and security" (Bowers 2015c).

in my youth   *says coyote*   everyone attended town hall meetings where speakers had to show links between events and words used to describe them   some abstractions feed the spirit if their quintessence is inspired by community enhancement   spiritual development

chet goes on   "The wisdom of the previous generations was encoded and re-enacted as part of the embodied experiences in the arts and ceremonies and the use of technologies and so on. They were closer to their environment not operating in an abstract world, although their mythopoetic narratives provided a conceptual overview of what their world was about" (Bowers 2015c).

says coyote   'aboutness   yes'   our deepest stories are not made up of words metaphors analogies or imaginary events   they are our cultural biographies our footprints and the dust that settles into them   they nourish us   challenge us   teach us   tell me chet   how the cultural commons relates to sustainability and respectful engagement across cultures

chet's ghost speaks   "The cultural commons are an aspect of culture that began with the first humans   that is the knowledge and skills mentoring relationships that were passed on from generation to generation and were not monetized" (Bowers, 2015d).

chet   *says raven*   you have some ideas about Freire and critical pedagogy in the area of emancipation

chet's gaze is steady   "The problem is that the traditions we are being emancipated from   are the traditions that limit our need to depend on expert systems and the marketplace" (Bowers, 2005, p. 175).

says coyote   you write that dominant culture is driven by techno-utopians who envision a world of machines replacing humans in the workplace   reducing lives to data for manipulation by governments multinational corporations and hackers (Bowers, 2016) sounds undemocratically brazen   *says raven*   I remember early cultural commons espousing values beyond monetary equivalence

chet continues   "It represented the activities that have a small environmental impact and that developed a sense of community, allowed its members to discover talents and skills .... that's passed on intergenerationally, games, language, social justice traditions, all of those activities that are less dependent upon a money economy and thus less dependent upon the industrial culture that is destroying the environment" (Bowers 2015a).

coyote pipes up   everything we do in the environment is or becomes environment   for we are it   not only are we environment what we buy is and becomes environment

raven flutters his wings   today community has eroded   the commons is not protected or honoured or valued as it was for millennia   individual greed is in the ascendant   people take more than their share if no-one is watching   steal what belongs to others   break every law not enforced   they have never known the idea of commons   nor cared   though at some base level it is there in cultural memory   underneath the remote control for the smart tv

chet says quietly   "Enclosure means turning what is shared intergenerationally and is part of community life into a product that can be integrated into the market system, or the enclosure occurs when there's a technology which undermines the importance of face-to-face inter-generational communication" (Bowers 2015d).

says coyote   academia   government and business misrepresent and destroy tradition because it gets in the way of their profits   power and   control   the rapidly escalating anthropocentric   monocultural virtual enclosing of the physical   intellectual   spiritual are machinations that are extinguishing the diversity of cultural commons locally and globally

chet continues   "Social justice traditions that were hard fought and won but are now threatened by being lost because of the digital revolution and learning from the printed word that appears on a computer screen. With the ecological crisis now forcing us to reassess this notion of progress, the cultural commons represent one of the areas in which we might recognize a possibility of carrying forward without further degrading the environment" (Bowers 2015d).

says coyote   "An eco-justice pedagogy is also based on root metaphors, but ones that have long been under attack by the idealogues of the Industrial Revolution and the Western Enlightenment project" (Bowers, 2002, p. 22)   who further augment this dismissal and assault of non-Western knowings and practices through their prioritizing of digital learning in current education reforms

says raven   'old' often means   'replace'   but 'new' means replace too because everything is made to last a short time before breaking down

and you can't buy parts or even tools or fix-it manuals so that the .001%
can increase on their tens of billions of dollars of personal wealth   making everything into an anachronism   planned obsolescence   out-of-date before it is even marketed because the next dozen generations of
'upgrades' are not only prototyped now   but market ready   landfill
ready   more   garbage   environmental   destruction   toxins   pollution   from   railroads   ships   planes   trucks   delivering   consumer
goods industrial machines to make more of what no-one needs but
everyone is trained to want   the excuse is always the same   economics   meanwhile   cultural commons gets left in the dust

the ghost of chet in profile turns to fully face the reader   viewer   listener   "The
Amish, Hutterites and other Anabaptist groups, they're carrying
forward the cultural commons within their religious framework....
The difference though is that the person learning to play an instrument and perform in a musical group, the person who is studying
dance, the person who is learning to share recipes within their
ethnic group, the development of personal skills   the ability to
write   [pause and a one and a two]   the cultural commons are current, and as students interact with the people carrying forward their
cultural commons, whether it's in weaving or painting or music or
storytelling or whatever it is, they're learning the values and sharing
characteristics of today" (Bowers 2015a).

coyote jumps in   it's a kind of "ecological intelligence that takes into account
the interacting patterns, ranging from how behaviors ripple through
the field of social relationships in ways that introduce changes that
are ignored by non-ecological thinking to how an individual's actions
introduce changes in the energy flows and alter the patterns of interdependence within natural systems" (Bowers, 2010, p. 24).

says raven   good on them but they've always lived that way   which is even
better   I've lived in cultural commons my whole life but have never
thought to designate it as being either 'culture' or 'commons'   it's
how we are taught to act in groups when living in the village

in my day   *says coyote*   ecological literacy meant you learned to read
the natural world you were part of   and of   which people from
mainstream   western   cultures   are   willfully   illiterate   (pause)
because   they see themselves outside of nature   chet says they see
nature as part of the technology-economic-market system and put
all their faith in science (Bowers, 1996)

the ghost   crosses the foggy stage   "The important thing about the cultural
commons is that it reflects an ecological form of intelligence where the
present activities very much are based on the past" (Bowers, 2015a).

coyote asks   can it make any difference to any but the already converted
and those loitering on the margins   this talking about the living
earth   honouring and respecting it   when industry and the great
unaware self-blinding consuming public go on ceaselessly to destroy

it every moment of their lives?   they recruit their children   their partners and parents   to help them   how do you reach people who don't care and might never care?   they will simply live in their heads   their appetites   their manifold and willful ignorance   as they have been formally educated to do   they will continue to buy the latest technology fad   application or gas guzzler suv until they drop dead   it will be only then that they will give something back

says the ghost looking at a giant screen of thousands of people not interacting except with cellphones and ipads "Technological determinism is also an expression of linguistic determinism ....We see in the educational process when progressive educators make the case that students should construct their own knowledge that they're autonomous individuals" (Bowers 2015d).

what I miss   *says coyote*   is silence and stillness   withdrawing from busyness and from our individual-as-collective addiction to refuse to acknowledge or pay attention to the world

the ghost smiles wanly   "What I see as the techno-fascist culture that's emerging... no one is aware that when traditions are lost or overturned you can't recover them so people just accept this notion, as you pointed out, of technological determinism, that we're powerless. We just go where the elites who develop these technologies and the corporates that gain economically from promoting them want, we just have to follow in line" (Bowers 2015a)."

says raven   chet bowers' eco-justice pedagogy might be a starting place   a place to work toward a green world as well as a just world   this will require letting go of the progress narrative of modernity to engage *ecojustice pedagogies of the land*   valuing Indigenous and other/ed knowings and practices as equivalent epistemologies   valuing orality and narrativity as equivalent ways of teaching/learning/knowing   acknowledging   the   human/non-human/more-than-human   intelligences   agencies   and interdependencies   this requires joining hands in solidarity   action and activism   with those locally and worldwide whose lives and lands have been destroyed by global capitalism and resource extraction   listening and dancing to the sounds and silences that have been on this land for millennia of millennia

adds coyote   I second that emotion (Robinson & Cleveland, 1967).

coyote   raven and ghost embrace

kukwstum'c   nia:wen

References

Bowers, C. (2016). *Reforming higher education: In an era of ecological crisis and growing digital insecurity*. Chambersburg, PA: Process Century Press.

Bowers, C. (2015a, Nov 28). The cultural commons: Alternatives to unsustainable living and technological domination. *Youtube*. Retrieved https://www.youtube.com/watch?v=7pdEZas7Su8

Bowers, C. (2015b, Oct 3). Part 3: Linguistic roots of the ecological crisis. *Youtube.* Retrieved from https://www.youtube.com/watch?v=42UCjOhnxJo

Bowers, C. (2015c, Sept 25). Part 2: Linguistic roots of the ecological crisis. *Youtube.* https://www.youtube.com/watch?v=ej6EG9fm70E

Bowers, C. (2015d, Sept 14). Part 1: Linguistic roots of the ecological crisis. *Youtube.* https://www.youtube.com/watch?v=O0YSPtPnNio

Bowers, C. A. (2013). The role of environmental education in resisting the global forces undermining what remains of indigenous traditions of self sufficiency and mutual support. In A. Kulnieks, D. R. Longboat, & K. Young (Eds.), *Contemporary studies in environmental and indigenous pedagogies: A curricula of stories and place* (pp. 225–240). Boston, MA: Sense Publishers.

Bowers, C. (2010). Plenary address: The challenge of making the transition from individual to ecological intelligence in an era of global warming. *Proceedings of the Media Ecology Association, 11,* 21–29.

Bowers, C. A. (2005). Afterward. In C. A. Bowers & F. Apffel-Marglin (Eds.), *Rethinking Friere: Globalizaiton and the ecological crisis* (pp. 149–188). Mahwah, NJ: Lawrence Erlbaum Associates, Publishers.

Bowers, C. A. (2002). Toward an eco-justice pedagogy. *Environmental Education Research, 8*(1), 21–34.

Bowers, C. A. (1996). The cultural dimensions of ecological literacy. *Journal of Environmental Education, 27*(2), 5-10.

Cabaret Lyrics. (n.d.). Lyrics.com. Retrieved from https://www.lyrics.com/lyric/1587829/Liza+Minnelli

Macfarlane, R., & Morris, J. (2018). *The lost words.* Toronto, ON: House of Anansi Press.

Robinson, S., & Cleveland, A. (1967). I second that emotion. Retrieved from https://www.songfacts.com/facts/smokey-robinson-the-miracles/i-second-that-emotion

Shakespeare, W. (1603/1623). Hamlet. Retrieved from http://shakespeare.mit.edu/hamlet/index.html

11 Developing Literacies through Place-Based Poetic Inquiry

A Curriculum of Movement, Travel, and Writing

Andrejs Kulnieks

When I think about ideas that Chet Bowers asked me to consider throughout my graduate studies and during my first years of teaching courses in Northern Ontario, I realize that many of our discussions explored the idea of how language and place intertwine. Bowers (2013) writes: Words have a history and as metaphors they carry forward earlier ways of thinking that were not informed by an awareness of environmental limits…" (p. 239). For me, and many other ecologically-minded thinkers, finding language to explore our relationships with the places that we grow to know, is the key to exploring and beginning to understand ecojustice education. In this chapter I consider how poetic inquiry helps me develop a deeper awareness of how language shapes my thinking about the word around me. I consider how learners can develop a deeper understanding of intact-ecosystems through returning and engaging with environmental learning in a particular place over a course of time.

As important as it is to engage with local learning, it is also important to go beyond familiar places to apply what is learned through communal activities, like food growing and harvesting, to explore and to create opportunities to share intergenerational knowledge. I document my own poetic inquiry though travel-writing as I journey northwest toward the Saskatchewan to work with curriculum theorists at the University of Saskatchewan. It is also the institution where Chet began his academic career. As I reflect on my journey of learning, I consider how going into unfamiliar spaces helps me to better understand the richness that the gift of returning to familiar spaces brings. Relationships with places along the journey develop both by returning to perennial locations and creating and engaging with life-stories, become a space for learning that move us beyond institutionally constructed spaces. I consider how the application of local knowledge about food can shape how poetic inquiry can help learners develop their relationships with the earth.

A Curriculum of Movement and Writing

As I write poetry, I also theorize a methodology of how ecologically inspired poetic inquiry can help learners to deepen our relationship with the places through developing a connection with some of the plants and animals that live there. I consider David Orr's notion that all education is environmental education, and how movement beyond the classroom makes this possible. As educators that are teaching learners about the importance of going outdoors, we often need to provoke conversations about the where to find clothing that includes an understanding of the value of thrift, as well as an appreciation for some of the skills that we need to enjoy our surroundings. How do we inspire pedagogies that re-consider technologies that are continually being sold to us as if they are the "necessities of life." Often, we are coerced to buy things as if we really need them to survive and find our place in the world. I consider Michael J. Lanoo's (2010) argument that the beginnings of environmentalism can be traced to movement between urban areas and non-urban spaces:

> Shacks, cabins, and their ilk, including shanties, huts, cottages, trailers, and even some laboratory buildings, reflect the architecture of a country founded in frontier. Shacks are outposts in wilderness. Shacks are places for freethinkers, people who want to get away from it all. (p. 4)

Shacks and spaces to write and congregate are a very important aspect of becoming familiar with in-tact ecosystems, as writers like Joseph Campbell, Aldo Leopold, John Muir, Ed Ricketts, John Steinbeck, Henry Thoreau, all spent a good time getting to know their local environments. Part of their inspiration came from their desire to be away from humanly reconstructed environments.

In *Wanderlust: A history of walking*, Rebecca Solnit (2000) outlines some of the battles that were fought so that there would be common areas that people could visit and move through. She wrote:

> Dorothy and William Wordsworth walking together through the Pennines just before the nineteenth century began seem lonely figures, choosing an unpopular activity in an unpopulated countryside, and John Muir tramping across Yosemite and the Sierra Nevada in the decades after his arrival in California in 1868 seems part of a tradition of solitary wondering, pursuing the aesthetic while those all around pursued the utilitarian. (p. 149)

One of the ways to develop a deeper understanding of the places we move to is through developing an understanding that it is important to return to the outdoors? Making a conscious effort to return to where you are in the future helps to instill the idea that you treat the places that you visit with respect. Often, we don't understand the importance of those paths until you have been there over a course of years. It is important to consider where those paths came from. Who made them? I often think about how they can still be there so many years after they were made. Do the forest animals maintain them?

As educators, we can bring attention to the importance of place through stories about particular places. Keith Basso (1996) describes the importance of understanding places and ancient place names remembered by knowledge of what has taken place at those places. He writes:

> For it is simply not the case, as some phenomenologists and growing numbers of nature writers would have us believe, that relationships to places are lived exclusively or predominantly in contemplative moments of social isolation. On the contrary, relationships to places are lived most often in the company of other people, and it is one these communal occasions—when places are sensed *together*—when native views of the physical world become accessible to strangers. (p. 109)

Basso goes on to explain:

> Relationships to places may also find expression through the agencies of myth, prayer, music, dance, art, architecture, and, in many communities, recurrent forms of religious and political ritual.
> (Ibid.)

The journey and experience of getting somewhere is often as important as arriving there. Part of a journey takes place before even going somewhere. How we travel often has a good deal to do with where we are in our lives and the opportunities available to us. Journeys often help us to recognize what is important. These start with at least a semblance of a plan. It is important to take time to consider what we know about the places where we decide to travel, and what we think we might learn about them in the process, as well as what we would like to learn through our travel experience.

Poetic Inquiry provides a space for me to reinterpret and engage with language through a reinterpretation of my stories. When we ask students to engage with their environments through movement and writing, we create an opportunity to consider what is important for us. Bowers (2002) writes:

> The difficulty in recognizing how thought and even material expressions of culture are based on root metaphors that reproduce

past forms of cultural intelligence and moral norms can be seen in how the supposedly most rationally capable members of society were unable to recognize the many expressions of patriarchy within the relationships and curricula of the university... When root metaphors are not recognized they largely dictate which analogs will be used to understand new phenomena. (pp. 22–23)

One of the ways that the Oxford English Dictionary defines analog is: "A thing which (or occasionally person who) is analogous to another; a parallel, an equivalent."

It is important for educators interested in ecojustice education to consider how learners can develop a deeper understanding of how language can shape our understandings about the world around us. Rather than thinking about language learning as the memorization of terminology, ecojustice education seeks to help learners expand their understandings about the world through what Bowers outlines as traditional and modern technologies:

An ecojustice pedagogy contributes to self-limitation for the sake of future generations when it helps students to recognize and participate in non-commoditized activities of community (Ibid. p. 32). Learning to enjoy the possibilities that, walking from place to place is one such activity. Walking old paths can help learners to consider the deep histories that places hold. It is important for us to develop habits that involve movement.

Stretching the body and learning the joy of maintaining health is an integral aspect of maintaining a high quality of life. If we don't practice and theorize healthy ways of movement, they are often ignored or replaced by sedentary activities like using devices. The following poem demonstrates how I attempt to motivate myself to move as I consider the importance of knowing the places that I live. It also helps me understand the importance of remembering what I once thought was important. I also employ the process of writing this poem to think through how my thoughts and experience with and about family and friends have changed over a course of time.

Retracing steps

walk paths to where you once were
you know where they are
you don't need to know exactly where to go
language reminds you

the words don't have to be exactly the same as everyone else's
recited verses built on scaffolding
where blueberries are not found at this time of year
complexity surrounds us

try writing a poem in one draft without mistakes
expectation of perfection dreamed
writing to measure cleverness or whit
or thinking through what you need to consider

stretch muscles to remember you have them
and they need you to use them or the will not be there
when you ask for their help
it feels good to remember to move

as you write your environmental autobiography to understand
how language changes thoughts in mind
you sway between the past and present
the world circles around you

someone explains, you might consider your behavior
revolving around the person in discussion to enhance the
hyperbole
the world does not revolve around you as the movement
of fingers and hands suggests
you need the world but it may be better off without you
if you keep going the way that you were

before you shift critical thoughts
moments pass you by
thankful for and that your body is good to you
so be good to your body

touch the sky as if you were swimming
move as if you were climbing a mountain
because you may find yourself staring at one
that you might want to see from a different point of view
(Kulnieks, 2021)

As part of my process of poetic inquiry, I question what happens if I add a photograph after writing the poem to revisit what I am thinking about movement. The meanings blend as do the subjects in this chapter. As I interrogate representations about what the words and images mean, I also search the photograph for other plants that I collect for tea. This reader may be thinking what plants do I know in the photograph?

The photograph and the meanings contained therein also stand on their own. Questions arise when I introduce the photograph. When was it taken? Who took the photo? Where was it taken? What is beyond the photograph? Something shifts in the reader's mind? They might ask questions like Where do I find berries like that? When am I going to pick blueberries? What do those blueberries taste like compared to

Figure 11.1 Wild blueberries
Source: Vectirele, 2021

store bought berries?" This too is part of how we develop an interest about the ecology of place.

Much of the time we might not know the history of the paths walked without accessing these histories. Jack Weatherford (1988) writes: "The present road and highway system, railroad network, and even the canals of the United States and other American nations largely follow Indian trails and roads" (p. 247). Getting beyond where we need to buy something to participate in activities is important to consider with an increasing amount of consumerism permeating every aspect of our lives. The language of education is increasing being transformed. The transformation that takes place through the act of moving from place to place helps the learner become in tune with the world around them.

Wendell Berry's agrarian essays help us to consider how movement and the consumption of fossil fuels has become an inescapable aspect of our day-to-day lives. The food that we eat is often not locally grown unless we take on the responsibility of how we live. Many find that the cost for buying local and organically grown food is not possible. Although it is becoming increasingly difficult to buy all the food that we need from local sources, travel writing becomes a space for considering how we live from day to day, and how we can better understand the world in which we live, and to develop a deeper relationship with the Earth.

Some of my fondest memories with Chet Bowers are from the ecojustice conferences and retreats that I attended. Often travel is messy in terms of border-crossings, and tiredness is one of the inevitable results that accompanies this form of movement. I remember a particular conference in Detroit where the key theme of the conference was "reclaiming the commons." As we toured the commons, Chet had clearly done a great deal of research and knew the city well. As we drove around the town, I was in the back of the van with Chet. He inspired me to think beyond the trip itself, and to consider how I could help my students to reconsider their relationships with the places that they lived through understanding the city as a living history. As I collected my thoughts, he kept nudging me and commenting that I would not want to miss this and can you imagine …

Connecting these memories of walking through Detroit with Chet Bowers and other ecojustice educators reminds me of Bowers' work on a curriculum of the commons. He wrote:

> The curriculum should also help them recognize the patterns and activities within their own communities that are still largely based on face-to-face, intergenerational sharing of knowledge and skills. These non-commoditized aspects of family and community life might range from dinner conversations made possible by a more balanced use of such modern technologies as television and computers to the existence of community theatre and other performing arts, mentoring in the development of individual talents, gardening, chess and poetry clubs, sports, and community service activities.
>
> (Bowers, 2002, p. 31)

As I investigate how poetic inquiry can help us to become more in tune with the world around us, I explore how my own participation in these focal practices outlined by Borgmann (1992) as ways of reducing the impact of nanosecond technology. I began to reconsider how my excitement about practices like picking blueberries, raspberries and knowing which mushrooms I could eat was a deep part of my active participation in the web of life, but that is not enough to know about the web. It is important to inspire students to consider their own practices by recognizing how some of the practices that they are already engaging with are an essential aspect of developing a deep relationship with the places that they live. As I travel across Canada to my new home in the Prairies, I consider how getting to the University of Saskatchewan, is also part of the mind-map that became engrained in the theories that Chet would have been working through as he began his work at the academy.

Enacting an Ecojustice Curriculum: Developing a Relationship with the Food that We Eat

For me, moving to a new province meant movement from farmers, markets as well as my own garden and a reliance on the produce that is sold in stores along the route that my journey took. I begin to better understand what is as stake when a connection with trusted food sources is lost. There is a feeling that I am relying on others to care for my health and well-being. When we lose connection with where our food comes from, we also relinquish control over what goes into our bodies. As I work with colleagues, they help me understand the importance of healthy ways of eating. Eating well can take many forms, and can lead to eating too well. Part of eating well includes developing an understanding of the foods and chemicals that we take into our bodies. This starts with something that those of us who were lucky enough to have gardens to get at least some of our food learned a long time ago. There is no food that tastes as good as the food that we have grown or gathered. As young adults, we often forget what is healthy, and even what our bodies need. Food stores are full of foods that are not designed for local community health. Rather, they are designed for sales. For this reason, it is important for educators involved in systems of schooling to provide structures for community building to take place. Veronica Gaylie (2009) explains:

> Amidst the tensions, the stewards and I began to understand how a *Learning Garden* requires hands-on work, and how perhaps more importantly, it requires a conceptual philosophy, based on shared principles and a communal work ethic to guide knowledge, labour and inspiration. (pp. 77–78)

The key is, that the less processed the food that we eat is, the better. Sugar, sometimes referred to as "white death" is not part of a healthy diet, but very difficult to avoid. The best processing for food is that which we do to it before we eat it. This is not something that big companies do to make food last for years and to be good for years after it was produced. Honey is one of the few foods that arguably lasts for decades in the fridge, and maybe longer. I know this because my grandfather was a beekeeper and the family would keep honeycomb in our fridge. I kept the last packet for over 30 years, and it tasted almost the same way that it did when we packaged it. Eating it released memories of working and taking care of the bees. For anyone who engages with beekeeping knows, is a labour of love. It requires perseverance and commitment. You can't leave the hives on their own during winter without the odd checkup as the weather shifts. It also requires a community of people that help when the need arises, as well as a demand for the honey that is produced. Family businesses need to cultivate healthy relationships with each other and often, there isn't a great deal of profit. It's important to support the small family business if we want them to continue working in spite of the corporate model that has been taking over traditional ways of living. The following poem is inspired by discussions with Chet Bowers as I consider many of the workshops and discussions about the importance of community that he participated with and helped to develop.

A pedagogy of landscapes

Move beyond what you are asked to buy
About the machine that you are born into
Resist temptations to not know what you can be
look at the local as you resist resistance

spend time thinking about how you think about a topic
that you think might be important to consider
but found it difficult at the start
but increasingly important as you moved towards completion

write about what you learn to know
terms that you will not be able call or send notes
Thanks for being approachable
as you were when thoughts seemed misinformed

lead journeys into places that you do not know
spaces that will inspire you to learn how you need the gifts that
Earth provides
though it does not ask or need anything from you
you owe it some thought and care

accessing photographs from decades long passed.
You made time to write those texts well before their time
because you knew that it was more important
and more difficult than destruction.

You shared your time
Because you realized that your art was to engage thoughts
That you could do better than most
And your art was to find language to recreate thoughts already known

So that they would be relevant to those who would spend the time
to listen
Walk thoughts back and forth until they began to become clearer
To those who would rather dismiss them
But knew that there was more to learning than following solipsistic
paths

You brought your family discussions to the forefront of mind
That would help others see the reasons to study those words
That meant different things to different minds
Not to forgot that life is not eternal but thoughts might be

One of the roles of ecologically-minded educators is to provide information that can lead toward healthy ways of living. As I write the poem, I consider how focusing on a moment helps me to better understand that moment, as well as those around it. It helps me to remember how thoughts that were so important in the moment can shift over a course of time.

Health, Movement, and Learning through Active Participation

The community-enhancing traditions, or what should be understood as the gift economy of the cultural commons—that is, the wealth that strengthens communities and leads to personal discoveries and development of talents—are not what is represented in textbooks or a printed text read on a computer screen (Bowers 2016, p. 89).

Classrooms are important spaces for gathering to consider ideas and to develop linguistic skills understandings about a place or what Gee (1991) calls secondary discourse. Acquisition of knowledge can be explored in the classroom but requires an? but also about the importance of moving beyond the classroom to engage primary discourse with place.

We often tell students that they should not stare at a computer screen for long periods of time but as educators, we often fail to provide activities that can be used to combat the health implications that the sedentary lifestyles that systems of education promote. We need to consider how movement and travel become a space for combatting the sedentary nature of much of the work that we do. Sitting in a chair for many hours of the day is not something that comes naturally. Teachers ask students to sit for much of our classes and act as if we are surprised that our students wish to move. Bodies are designed for movement at every level of being. It is important to practice (and figure out) activities that help them stretch muscles. Movement is something that can inspire focus on sedentary activities.

Steps back toward the gift of movement

Move as if you would if you had to do everything
By yourself
Stretch as if you knew you had to do all that you can
With your body

There are many ways to create the scene
Stretch hands towards the sky as if to touch the clouds

It only takes a minute
to see where your neck can go

remember movements that you might need use
try them when your life doesn't depend on them
so that when it does
you have practiced

lubricate joints through movement
without oil
start from the top
move down

and once you have opened channels
to the universe
begin to work through
what your eyes can do

before to sit in front of the screen
after your sit in front of the screen
when you have sat down for a while
your boss will not make time

it is a request that can only be made
when we realize that it is not optional
if we want to be able to use the parts
for a long time

these activities are not written in stone
there are masters who know the slow movements
that lead to strength
but movement takes time.

(Kulnieks, 2021)

Writing through a methodology of poetic inquiry helps me to consider the bigger pictures. All the components are there, but the journey of learning requires participation and for these activities to be beneficial to as many people as possible, educators need to make time to inspire communities of learners that are willing to take the time to consider how important it is to take time to move around. Movement was a part of the daily lives of all of our ancestors and although the type of work that many of us do involves sitting and standing, we have to provide opportunities for our bodies to do more.

Travel Writing as a Return

One of the most important reasons for travelling is to return to places that we have been as we develop a closer relationship to those places as

Figure 11.2 Mirror reflection
Source: Kulnieks, 2021

we dig deeply into what we would like to know more about. Each year in the Latvian diaspora community, we try to keep our relationships alive by making certain events alive. Winter solstice is one where we congregate with family. Summer solstice is spent at one of the summer camp locations outside of the city and during the autumn season we try to meet on the Thanksgiving weekend with family. These weekends are connected with shifts in the seasons but clearly, the longest and the shortest day of the year are celebrations of the sun. Each year, our family celebrates this as a fishing weekend in northern Ontario. The following is a poem that documents that experience from my perspective. The following photo is also serves as a representation of the boat ride as well as a focusing moment for the poem being written.

The journey continues

Glide down the river to where I have been
Over forty years
At this time of year
To witness the rebirth of the season

Spring shoots move soil to the surface
Signaling the beginning of their growing season
Towards the sun
A dance of movement

Moss hides
under patches of ice
last moments of rest
aside the symphony of blackflies

travelers on this journey
who have already gone beyond the sun
are still with us
as we journey onward

the pike that feed us throughout the weekend
have been here
long before we have returned
they make the weekend magic

flake away from the bone
soaked in batter
crispy, light brown
as they return to the web of life

it is cold as the sun rises
but when the warmth visits us
the hum of the blackflies remind us
of what will be here soon

the waving grass is all that lets us know
what side is up
you will be with us the journey
even when camera lies

(Kulnieks, 2021)

The poem communicates with the photograph in real time. As I write
the poem I realize that I have only told a snapshot of what is taking
place. There is a great deal more to my story, but the story I tell is only
part of what the reader will interpret. As Louise Rosenblatt points out:

> The individual's private associations with a word may or may not
> agree with its connotations for the group, although these connota-
> tions must also be individually acquired.
>
> (Rosenblatt, 1994, p. 1060)

The reader will not interpret the poem as I have written it in my manuscript. Even the poem that I am citing for the purpose of discussion in
this essay may have changed by the time that this book is published, and
certainly by the time that it will be published elsewhere. I am hoping that
the poem helps to highlight the importance of returning to a particular
place over a course of time. The importance of this return is also true in
rereading and rewriting as Sumara (1995), among others describes.

Modeling Ecojustice Approaches to Research and Inquiry as Part of Curriculum

One of the key reasons to travel is to deepen understandings about the world around us. From time to time, I embark on a research journey that helps me better understand the world around me. One such journey was to visit Australia, as I embarked upon my doctoral research to compare the Latvian Scouting movement in Australia as well as in Canada (Kulnieks, 2009 Part of this research involved visiting a school in the Outback as I began thinking about how my relationship with the environment had developed thorough my years of public schooling, and how that would be different if I was living in non-urbanized environments.

Part of this research helped me to consider how place-based education could use the methodologies of movements like scouting and girl guides to help learners develop a better understanding of active citizenship. Geraldine Balzer describes the importance of understanding civic duty in one of her articles entitled: *Why go to Guatemala: International service learning and Canadian High school students.* In her research she considers the importance of providing opportunities for learners to go beyond their zones of comfort, to consider their lives from living with people from very different perspectives. By living in the same place people that have a close relationship with the land that gives them life, they also begin to share understandings about how vital it is to consider multiple ways of thinking. In particular, they had an opportunity to understand the gift that plants give us, and to be part of the process of experiencing the interconnection of living in a close relationship with the Earth. The unfamiliarity of being in landscapes that we are keen to learn about is also an opportunity to see things that we know with a different attitude which is one that can't take things for granted.

A big issue that all citizens of the Earth need to consider is how the way that we walk the Earth will impact future generations. Students can consider how their actions can contribute to positive changes that action, like attempting to stop plastic use, can promote. It is not an easy task. About a decade ago, I remember making a pledge with a group of like-minded researchers, to stop the use of plastic. The most important thing that I learned from this pledge was not that this was not possible, but rather, that through being aware of the evasive nature, creating awareness can inspire the spread of information that can begin to make a difference. Some of the biggest issues around plastic use include the time that it takes to break down, the convenience that plastic provides us with and the revenue that gets generated through production.

Here in Saskatoon, there are things being done. For example, not only is there a non-refundable environmental handling fee included in the price of using plastic, but there is a company called SORCAN where residents can return plastic for a refund. Milk cartons, jugs, cans and bottles all have a deposit fee on top of the non-refundable "environmental handling charge." In the following poem, I document my memory of a snowy day.

Bike-store interrogations

Is your lunch packed in reused materials
a few thoughts a week can make a difference
consider the effect that plastics will have
if they aren't going away

did you walk or glide through snow, wind or ice
or did you make it in your gas guzzling chariot
is it shiny and new
or did it take you long roads for a quarter of the century

did you think it through?
was it told to you
did you show up with civic transportation
with your coffee mug

did you mention to the cashier
that there was a reason that you brought your own cup
with a smile
or bypass the lines altogether

did you read the package
rewriting and interrogating ingredients
when you were you were unsure of the effects
and say no to some of the cravings that were tempting you

did you see a pollinator
and know what it was
gave it a lift on the page in the midst of your day
and let it continue to move outside

did you bike to work
or walk to the restaurant
go for a walk instead of watch a video
decide not to buy something that you have and could keep using

did you decide to plant a garden
buy something locally grown
go somewhere close for a vacation at a different time of year
to save the money (and fuel) for a flight?

if you were thinking through the ideas
letting them flow through your day
making others think about alternative choices

you may be attempting to become an even more ecologically minded citizen.

(Kulnieks, 2021)

Many educators often take it for granted that learners should know what they are doing in terms of recycling and other ecological imperatives. The truth is that many do not, and more importantly, need reminders to consider how simple actions, like sorting garbage and being mindful of what they buy, can make a difference in the future.

Implications toward an Ecojustice Education: Travel to Deepen Relationships with Place

For young people, this is particularly important. Teaching in the area of Geography, I have met many students who have a very poor understanding of the cities that they live in. I will discuss two of the projects that act as culminating activities for social science courses. In some cases, it is important for students to become familiar with the subway systems and public transportation systems that are in place. It is important for educators to design curricula that ask students to imagine that they develop a relationship with some of the public land that they can access in the city. This type of work asks students to consider some of the public spaces that are not developed to become more familiar with them through theoretically based activities. These spaces are what Bowers describes as "the commons." These may be places where students congregate to play sports or to "hang out" or to share intergenerational knowledge through family activities like picking berries or mushrooms, depending on the types of places that they have access to.

I question where the journey begins and ends. Photographs and words only represent a moment. When we look more deeply into the words, we find a more meaningful story. Beyond the snapshot from photo lens the world is moving. Fish are hunting for food, roots are slowly moving earth, and the pollinators are coming to life. It is important to consider how to get students away from the devices that have become part of their lives. This is not a new concern. Neil Postman (1985) wrote about the dangers of the televisions years ago. There is much to be learned along the paths that Chet has taken us. I miss those fireside chats as we attempted to unravel some of the complex web of ecological theories of language and place. There is always more to see and more to consider. The key is to provide spaces for to consider the way that language shapes our understandings about the world around us, and to develop strategies that will help combat the mechanistic nature of learning that is increasingly being dominated by corporate interests.

References

Basso, K. (1996) *Wisdom sits in places: Landscape and language learning among the Western Apache.* Albuquerque, NM: University of New Mexico Press.

Borgmann, A. (1992). *Crossing the postmodern divide.* Chicago, IL: The University of Chicago Press.

Bowers, C. (2016). *Digital detachment: How computer culture undermines democracy.* New York, NY: Routledge, Taylor and Francis Group.

Bowers, C. (2013). The role of environmental education in resisting the global forces undermining what remains of Indigenous traditions of self suffi-ciency and mutual support. In A. Kulnieks, D. Longboat, & K. Young (Eds.), *Contemporary studies in environmental and Indigenous pedagogies: A curricula of sto-ries and place.* Rotterdam, the Netherlands: Sense Publishers.

Bowers, C. (2002). Toward an eco-justice pedagogy. *Environmental Education Research, 8*(1), 2.

Gaylie, V. (2009). *The learning garden: Ecology, teaching, and transformation.* New York, NY: Peter Lang.

Kulnieks, A. (2021). *Walking through thoughts: Developing a relationship with places.* Unpublished poetry manuscript, University of Saskatchewan

Kulnieks, A. (2009). *Ecopoetics and the Epistemology of Landscape: Interpreting Indigenous and Latvian Ancestral Ontologies.* Unpublished Dissertation, York University (2009).

Lannoo, J. (2010). *Leopold's shack and Ricketts's lab.* Los Angeles, CA: University of California Press.

Postman, N. (1985). *Amusing ourselves to death: Public discourse in the age of show business.* Toronto, Canada: Penguin Books.

Rosenblatt, L. (1994). The transactional theory of reading and writing. In *Theoretical models and processes of reading* (pp. 1057–1092). Newark, Delaware: International Reading Association.

Solnit, R. (2000). *Wanderlust: A history of walking.* New York, NY: Penguin Books.

Sumara, D. (1995). Response to reading as a focal practice. *English Quarterly, 28*(1), 18–26.

Vectirele, Z. (2021). *Images on the move: Capturing the journey* (Unpublished manuscript).

Weatherford, J. (1988). *Indian givers: How the Indians of the Americas transformed the world.* Random House Publishing.

12 Ecocritical Pedagogies

(Re)Imagining Education for Diversity, Democracy, and Sustainability as Eco-Justice Curriculum

John Lupinacci

The first two decades of the 21st century have been an era in which many people have found themselves facing basic issues of survival on Earth. Growing evidence suggests that heightened racism, sexism, and a fearful intolerance of queer and differently abled identities are inextricable from a white supremacist, hetero-patriarchal, ableist and industrialized consumer culture. Through a rise in tyrannical governments and a global pandemic, 2020 has put into clear focus a culture that is not sustainable for human life on the planet. Amidst the grim reality that this shift has taken a brutal loss of human and more-than-human life and a gross denial of civil liberties, the topics of social justice and sustainability have been brought into serious consideration. In addition, more and more of the world's people are taking action to resist the spread of such socially unjust and unsustainable habits of living and demanding change from governments. As teacher educators, we are without question closing the second decade of the 21st century immersed in a range of interrelated social and political movements that are calling for our diverse efforts in supporting a collective saying "no" to regimes of tyrannical powers and "yes" to each other and to a multitude of sustainable and socially just ways of living. It is imperative in a world fighting to do more than survive that teacher educators take seriously this moment to radically reconsider not only the purpose that schools have historically served in society but also to critically imagining how the purpose of diverse schools might serve a different kind of society.

Arguing for an ecojustice-oriented education that informs students of the politics of not only environmental degradation but also of the inseparable social injustices suffered in society related to some deeply rooted cultural assumptions in dominant human communities. As Bowers (2002) wrote:

> The implications for an eco-justice pedagogy include providing a critical understanding of the deep cultural assumptions that under-lie the industrial and consumer dependent form of culture as well

as an understanding of how the languaging patterns of different western cultures create the individual psychology that accepts consumer dependency and environmental degradation as a necessary trade-off for achieving personal conveniences and material success. (p. 30)

It is highly problematic and ultimately undermining to social justice efforts for environmental activism—especially in education—to separate environmental degradation and exploitation from social injustices entirely. While it may be helpful in some cases for researchers to specialize on single issues in specific reports and studies like rising sea levels, it is not helpful for scholar activist educators to see these studies in competition. More and more we are seeing that intersectionality and the interdependence of these social and environmental problems are being analyzed systemically and at their root cultural causes. Bowers (2002) outlined three aspects of an ecojustice pedagogy as environmental racism and class discrimination, recovery of non-commodified aspects of community, and responsibility to future generations. While I find all three incredibly important to my work as an eco-critical scholar-activist educator, I find the second and third points to be my primary anchor, and I take the first aspect as an example of connecting racism with social class but in need of thinking more inclusively about the complex intersections of race, class, gender, sexuality, ability, and even species as part of an ecojustice pedagogy. Having been able to work closely with Bowers and especially with others like Rebecca Martusewicz and Jeff Edmondson and their students, I am reminded that an ecocritical perspective does not mean following one line of critique or analysis but rather a responsibility to addressing unjust suffering in our communities as well as societal problems through which oppression is rooted in concepts like patriarchy, white supremacy, and capitalism. Considering such an important responsibility and the current political moment for the planet, I am drawn to beginning this chapter focusing on how Bower's work in ecojustice pedagogies influences me currently to consider possibilities for ecocritical pedagogies in teacher education.

Building on a legacy of civil rights activism in the United States and around the world, Black Lives Matter (BLM) emerged from #BlackLivesMatter in 2013 after the acquittal of George Zimmerman— the man who shot and killed a Black teenager, Trayvon Martin, in 2012. The BLM Movement grew as protests emerged in response to the deaths of Michael Brown, Eric Garner, and Sandra Bland. By 2020, the kind of state sanctioned violence against Black and Brown people had boiled over. Amidst a planetary pause that further illuminated systemic inequities brought about by COVID-19, the murders of Ahmaud Arbery, Breonna Taylor, Elijah McClain, and George Floyd called for extreme action in defense of not only Black lives but also of the issues outlined by the BLM movement. BLM affirms that racial justice is intertwined with queer rights, climate justice, and eliminating sexism and poverty for all suffering people by illuminating the ways in which white supremacy works to undergird and rationalize injustice. BLM in 2020 became an even stronger resurgence of civil rights activism and did so with attention to the importance of solidarity against all the ways lives, especially Black lives are systematically targeted around the world for their demise and often death. Similar in 2016, the #NoDAPL drew massive attention and support to the Standing Rock Sioux's grassroots resistance to the planned and approved construction of the Dakota Access Pipeline—a 1,172 mile long (approx. 1,886 km) underground oil pipeline in the United States. The Indigenous centered protest brought to the forefront of social activism the long-standing importance of water as life sustaining and sacred. In 2016, the world saw "Water Protectors" standing in the way of the bulldozing of sacred land attacked by police dogs and sprayed with water cannons in below freezing temperatures.

Our current moment culminates from consistent civil unrest and protest often directed at the Trump administration, and the unjust policing of borders and bodies—including unjust imprisonment, incarceration, and murder, combined with over 2.5 million COVID-19 confirmed cases and 126,000 deaths in the United States as of the end of June in 2020 (Center for Disease Control, 2020). Yet, at the same time, tens of thousands of people protested in support of BLM in major cities around the world (Braslow, 2020; Parker, Manasce Horowitz, & Anderson, 2020). These protesters are risking health and taking action to say enough is enough, supporting the BLM movement and demanding political change. They are inarguably defenders of democracy. Protestors, and especially youth, around the globe are stepping up in significant ways. Frustrated with the injustices of mass shootings, racial inequality, intolerance for diversity and continued disregard for human rights and the health of the planet, the youth voice from Standing Rock and BLM to the Parkland Youth and Climate Change Strikes has been loudly heard around the planet. For people in the United States, and around the world, these protests are a statement of solidarity in tumultuous times and evidence that as human beings we will take whatever action necessary toward ensuring that diversity is valued democratically together with human rights and sustainability. We are living in dangerous times while protesters are courageously pushing for radical change toward social justice and sustainability now.

As a teacher educator with a vested interest in democracy, human rights and sustainability, our challenge, amplified by BLM, is to pursue an educational revolution in this historic time will urge educational leaders to strongly consider the alternatives to simply returning our schools and communities to the status quo. Furthermore, in conjunction with the efforts of leaders, teacher educators and teachers increasingly recognize that it is time to take direct action specific to teacher education. Lessons learned from a global pandemic and the current fight against racism over the past year demonstrate calls for a path which deviates from the current practices of teacher education and offers a rare opportunity for systemic change—one with which our nation is not unfamiliar.

The ideas that I share in this chapter are neither new nor ought they be attributed to originating with, or belonging to, any single author or academic camp. My hope is that they serve as powerful public testimony that there have been and are teacher educators responsive to a diversity of world needs, wants, and demands. As the research on educational trends in teacher education (Cochran-Smith, Feiman-Nemser, McIntyre, & Demers, 2008) and the impending crises we face in the next half-century (Stengers, 2015; DeLeon, 2019) clearly demonstrate, a refusal to understand and embrace mutuality and interdependence is woven throughout the interconnected hardships of social suffering and environmental degradation (Lupinacci, Parkins-Happel, & Turner, 2018a; Lupinacci, Happel-Parkins, & Ward Lupinacci, 2018b). This refusal is embedded in a conceptual framework based on a system of exploitation and violence—a lens that serves as the dominant shaping force regarding what it means to be a teacher. In this chapter I will argue that a status quo educator is one self-focused on the individual in a socially constructed value-hierarchy of a teacher as framed by patriarchy and white culture that acts as a superior in control of students. I further argue that this conception must be examined and critiqued for its limitations in favor of efforts toward an *ecocritical* pedagogical (Lupinacci, Happel-Parkins, & Turner, 2019) approach focused on the health and well-being of a broader human and more-than-human community, a community in which the educator is a mediator in rethinking the foundational roots of unjust social suffering and environmental degradation together with students engaging in critically examining and changing society. In confronting this stark contrast, I emphasize Bowers work in particular to educators as they work toward an ecojustice curriculum. Finally, I will argue that such work is a call for teachers to take action as responsible agents in shaping present and future society. Or in the sense that Bowers (2002) suggests that teachers take responsibility for future generations.

Teachers are critical leaders, and especially as teacher educators, we have a responsibility to examine and address how schools create, support, and sustain the violence of social suffering and environmental degradation. When leaders are faced with such challenges, we must be willing to inquire into the ways that current forms of exploitation are rationalized, justified, and/or ignored. In accordance with this inquiry, the purpose of this chapter is to propose a possibility for (re)imagining teachers as part of a movement toward horizontal partnerships in our communities by presenting an eco-critical conceptual framework for teacher education. Calling for a particular kind of teacher leadership supportive of social justice and sustainability, I share an introduction to such a framework as it contributes toward ecocritical pedagogies. I conclude this chapter by calling for a particular kind of teacher leadership supportive of social justice and sustainability. Such leadership works through direct action against the status quo that created conditions in which teachers confront a global pandemic and continued violence against oppressed peoples, animals, land, waterways, oceans, and skies.

Responding to the systemic violence and exploitation perpetuated by the current dominant social, economic, and environmental contexts in North America (similar to other nation states embedded in Western industrial culture), ecocritical educators examine and address how it is that schools in Western industrial societies create, support, and sustain the habits of mind that rationalize, justify, and (re)produce unjust social suffering *and* devastating degrees of environmental degradation. It is important that this work include a critique of white supremacy and tyrannical forms of government as ecocritical frameworks should not be confused with an ecofascist argument which utilizes climate change science to argue for exclusionary practices that rationalize authoritarianism and militarized actions to ensure continued unjust social superiorities. I want to be clear that action taken to address climate change or environmental degradation and domination as separate from racism, sexism, ableism, classism, and heteronormativity is part of the problem. Ecocritical work in teacher education asserts that addressing social and environmental justice as inseparable is paramount to future educators unlearning current dominant roles teachers and schools play in supporting or undermining the importance of diversity, social justice, and sustainability. When faced with such a challenge, ecocritical teacher educators ask: *How is it that exploitation and domination is rationalized, justified, and/or (re)produced by schools and teachers?* Furthermore, ecocritical educators are committed to turning the critical lens inward and asking: *What can teacher educators, and teachers, do to teach in support of alternatives to Western industrial culture?* Or in the case of current politics in the United States, in support the BLM movement as inclusive of local health and sustainability. In an attempt to address these questions, I introduce an ecocritical pedagogy as one among many valued approaches to exploring the possibilities of diverse critical ecological perspectives in teacher education.

An Ecocritical Framework in Teacher Education

An ecocritical pedagogy, addresses how education is influenced by systems of exploitation and violence, systems which rely on a refusal to acknowledge and embrace mutuality and interdependence (Lupinacci & Happel-Parkins, 2016; Lupinacci et al., 2019). Ecocritical scholars in education use diverse critical lenses for addressing and rethinking dominant cultural frameworks, but certain principles remain at the center of this work (Lupinacci et al., 2018a). Specifically rooted in and yet pushing the boundaries of the critical tradition, teacher educators positioned within the ecocritical movement recognize that social and environmental justice are inseparable and inextricably linked, and that these injustices rely on the perpetuation of cultural habits—like domination, individualism, and consumerism. They also acknowledge that such habits are deeply rooted in value-hierarchized, often dualistic, social thought—dualisms such as culture/nature, male/female, mind/body, master/slave and reason/emotion—that influence collective attitudes and unexamined behaviors. To dismantle such injustices, these scholar-activist educators analyze the culturally constituted value hierarchies that our society is reproducing. This approach also includes exploring diverse knowledges and ways of recognizing and understanding difference that move beyond the limitations of Eurocentric (or Western industrial) thought and the complex tensions, double-binds, and even contradictions that exist within our modern cultures seeking short-term solutions. For example, this means acting and acknowledging the importance diverse Indigenous wisdom and relationships to land and ancestry specific to place and key to decolonial moves in our schools and communities. In another example, we often find ourselves consuming commodities in an effort to improve our lives and even support so called "green" practices, but our relationship to these goods, to the people who create them, and to the beings, who supply the materials, often remains invisible and unquestioned. Our methods of consumption and participation in society are deeply influenced by our culture's hierarchies of value. As such, ecocritical scholar-educators ask: *How are our social practices and relationships dependent on privilege and exploitation, and in what ways might these practices and relationships be restructured to be more inclusive without exploitation and exclusion?* In other words: *How might we restructure practices and relationships to support feminist, anti-racist, decolonial classrooms and communities?*

Ecocritical pedagogies in teacher education strive to engage teachers in identifying and critically examining the role that education both plays, and ought to play, in transitioning toward supporting diverse, socially just, and sustainable communities. Drawing from an EcoJustice Education framework (Martusewicz, Edmundson, & Lupinacci, 2020) and what Martusewicz (2019) defines as a pedagogy of responsibility that stems from the growing field of ecocritical work in social and cultural foundations of education. EcoJustice Education can be summarized by three tasks:

1 Learning to analyze the deep cultural roots of the social and ecological crises plaguing our world.

2 Working to identify the diverse cultural practices that encourage relationships of mutual care of both human communities and the natural world we depend on. We call this second task revitalizing the cultural and environmental commons.

3 Developing the capacity, skills, and imagination needed to recognize what are mostly unconscious ways of being that are harming the world, and to create solutions, not just for the future but now, in our present context.

(Martusewicz et al., 2020, p. 18)

Simply put, through an ecocritical framework, teacher educators work to support scholar-activist educators in recognizing two conflicting and foundationally different worldviews. On the one hand, are ecological worldviews of interdependence and interspecies equity, and on the other, a human-centered worldview informed by hierarchal capitalist, racist, and patriarchal relationships. Following the three primary tasks of an ecojustice education (EJE) framework, I argue that nested in the third task is important work for teachers to examine and identify how to teach or share skills, and habits of mind, that support socially just and environmentally sustainable communities (Lupinacci & Happel-Parkins, 2016). Simultaneously, this framework shapes research in teacher education that examines how those worldviews might be reconstituted—via education or ecocritical pedagogies—in ways that are local and in support of living systems.

Ecocritical pedagogies often include students and teachers examining how knowledge systems and the associated behaviors and mechanisms of human domination—in relationship to language, culture, and power—are culturally mediated, and how PreK-12 educators can play a role in reconstituting those understandings, behaviors, and mechanisms. Furthermore, with(in) ecocritical pedagogies, it is important to recognize that the role of teacher can be taken on by more-than-human members of any learning community. By highlighting how several of the dominant cultural belief systems, root assumptions, and narratives currently destroying the planet and prospects of peace are constructed and not simply "natural," educators can help students develop critical perspectives on these root beliefs. This undertaking also allows for the exploration of alternative belief systems and metaphors that facilitate different kinds of relationships with other people, other beings, and the land.

An ecocritical pedagogical framework also illuminates the systematic economic and political restructuring of lives that, while it might address immediate needs of human or environmental systems, ultimately may perpetuate unjust social suffering and extreme environmental degradation. For example, ecocritical pedagogies require moving beyond a simple critique of technological development that portrays technology, or humans, as the problem or the solution. Rather, on one hand being critical of the dominant systematic economic and political restructuring of lives in current hyper-consumerist Eurocentric cultures includes also recognizing how technologies like pace-makers and vaccines have been, and probably ought to be, valued contributions to addressing human suffering and understanding the impacts of pollution and the importance of remediation and of green energy. However, ecocritical pedagogies require that teachers and students also ask how these valued technologies may or may not be sustainable and equitable via renewable resources and fair labor. Ultimately, addressing unjust suffering in society is important, but it must also include an awareness and analysis of the impact of any proposed solution on several generations of both human and more-than-human communities. What has become commonplace over the past century, and extending into the current, is the intentional restructuring of relationships to control and commodify lives in order to maintain and manufacture markets. For example, food and water are life-sustaining elements necessary for supporting healthy communities. However, the relationships to these "resources" have been enclosed—that is, they have been monetized or understood as commodities to be earned and purchased. This iteration of capitalism as a "supply and demand" economic system predicated on exploitation works to enclose living systems and can be understood as the globalizing force to commodify and privatize that which is common and public. Critiquing this kind of enclosure means reimagining the sorts of relationships humans can have with the more-than-human world and with each other, rather than simply trying to make human behavior "less bad" by reigning in largely minor destructive practices.

In short, ecocritical pedagogies center student learning on recognizing the importance of examining intellectual, environmental, and cultural practices and traditions in regard to how they either support or undermine living systems together with whatever content is being taught. Whether examining discursive practices or economic structures while learning mathematics, language arts, science, or social studies, a key feature of ecocritical pedagogies is the recognition that human knowledge systems are culturally constructed, have consequences for all living beings, and can be re-imagined in transformative ways (Turner, 2015; Lupinacci & Ward Lupinacci, 2017). Another distinguishing aspect of ecocritical pedagogies is that, whatever the lesson or activity, students and teachers together address the powerful role that their culture plays in the development of themselves, their values, and their diverse relationships. Such a framework examines, explores, and proposes diverse and collaborative pedagogical projects that respond to current dominant belief systems, working to ensure that any responses are necessarily collaborative with diverse cultures in ways that are local, situational, and in support of decentralized living systems. In the classroom, this might include examining clips from news media to determine what root metaphors are at work in the way our culture communicates about the more-than-human-world. It might mean exploring personal relationships with nonhuman beings through art or creative writing. It might mean comparing historical writings about race, gender, and species. Or it might mean using interdisciplinary knowledge to investigate a local issue of environmental justice.

A primary premise in ecocritical work in teacher education that differentiates this approach from most other critical frameworks is the explicit recognition of the entanglement of human supremacy and Eurocentric culture–that is, the embedded worldviews and belief systems originating in Western traditions of thought, which have colonized much of the world. Ecocritical educators assert that situated at the root of social and ecological injustice is a fundamental—and problematic— assumption that humans, as a species, are understood (or self-identify) as superior to and somehow separate from all other living beings and non-living things. Thus, guiding ecocritical pedagogies is the understanding that the manifestation of a human-supremacist worldview is culturally constructed and inextricable from current Eurocentric (patriarchal, classist, ableist, speciesist, and racist) industrial dominant cultural assumptions about race, class, gender, ability, and age. A foundational tenet in ecocritical work is that cultural habits of mind in dominant Eurocentric industrial culture are based on a system of human-supremacy—stemming from anthropocentrism—and that such a perspective is ubiquitous throughout dominant colonial culture and informs how we as humans in such a culture learn to interpret and assign value to differences.

Ecocritical work continues and pushes the boundaries of the work of social justice scholar-activists within critical education spaces by making clear the connections between human-supremacy, patriarchy, racism, and other forms of domination like ableism, ageism, and classism. Ecocritical scholars understand that each of these value-hierarchized structures of domination mutually reinforce one another while they obscure the fundamental interdependence and interrelationship of all beings. These hierarchies inform dominant cultural assumptions in Western industrial culture, and they are all based on normalized logics of domination, which means that they are inseparable and intersect in complex and contextualized ways. For example, the water crisis in Flint, Michigan or the Water Protectors in Standing Rock protesting the oil pipeline illustrates the complex intersections of anthropocentrism, racism, and capitalism.

Teaching for a Socially Just and Sustainable Future

The more that educators engage ecocritical pedagogies which encourage the recognition of and resistance to all forms of domination, the more potential there is for educational experiences to foster spaces where teachers and students learn together to recognize the harmful assumptions and actions that undergird social and ecological injustice. While on one hand we admire, value, and are firm supporters of a shared commitment to respond to the undeniable atrocities that we—as humans—enact on one another, these atrocities are inextricably connected to the cruelties we perpetuate against non-human animals and the environment. None of these atrocities occur in isolation—and, as outlined in this chapter, they echo the call from Indigenous educators and leaders in social movements like the Standing Rock, the Youth Climate Protests, the Women's Revolution in Rojava, and BLM. They are all interconnected. To confront human supremacy together with other forms of supremacy in education, it is paramount that educators work as committed allies to those suffering while challenging and confronting the systemic roots of oppression on our respective fronts both to address immediate ways human communities are suffering and to center inquiry on interdependencies and the interconnections among rights and needs. In other words, we all have a responsibility—many of us as privileged members of society—to support those, including our more-than-human kin, who are suffering unjustly, in whatever capacity we can while recognizing that solutions to end such suffering must not include an unsustainable exploitation of one another and the Earth's finite resources.

Responding to the enclosures of schooling by connecting the systemic roots of anthropocentrism to racism, sexism, classism, and ableism requires attention to be turned toward the difficult necessities for cultural change. As an ecocritical activist–scholar educator, I believe that if the enactors of dominant Western industrial culture do not rethink the cultural framework rooted deeply in our language, then we are destined to re-create and perpetuate many of the problematic relationships that we as radical educators often set out to change. Inspired by movements to address unjust suffering for our human and more-than-human kin, I am suggesting that we listen to and learn from social movements and the activist leaders like the leaders of the BLM movement, The Women's Revolution in Rojava, or the Water Protectors on occupied Native territories together with taking some actual steps in our studies and enactments of pedagogies in our classrooms toward cultural change. These suggestions with which I am concluding are aimed at supporting a paradigm shift from rational, mechanized, and human-centered, white-male-heteronormative thinking to discourses that are local, situational, and supportive of all living systems.

As an ecocritical scholar-activist educator deeply suspicious of educational mandates, I cannot suggest or outline any practical steps without including the importance of a fundamental shift in the very common assumptions, cultural relations, and traditions that define schooling. Drawing from my experiences with ecocritical pedagogies within teacher education, I will share steps that can be used to help students value pedagogical practices that work together with activist networks to resist what is outlined in this chapter as the logics of a particular kind of domination in favor of recognizing and valuing differences. These pedagogical suggestions are in support of diverse, decentralized communities where decisions are made with close consideration for all those species and groups directly impacted by the decision. Although I recognize the importance of localized responses and understandings, below are some preliminary suggestions for how teachers, and teacher educators, might begin to utilize an ecocritical framework in their lives and classrooms.

- Engage in teaching and learning that explores diverse projects to rethink the dominant assumptions influencing how we, as humans, construct meaning and thus how we learn to relate to each other and the more-than-human world. Further, make the commitment to critically and ethically examine how, as teachers, we individually and collectively understand educating, organizing, and taking action toward supporting healthy communities, communities that include the intrinsic value of recognizing, respecting, and representing the needs of all beings to belong to and live in peace within an ecological system. For example, teachers will need to critically engage in questioning how we language our world. Asking: *What does it mean to refer to natural gas and oil reserves as "natural resources?" What are we ignoring when we commodify the environment in this way?* Asking also: *What language—and in particular metaphors—further perpetuates social inequalities and undermines human rights in our communities?* As future teachers, also think about how we might frame lessons that include ecocritical essential questions such as: *What does it mean to be human beings in our diverse communities of life? Who/what benefits and who/what suffers? What are the deep cultural assumptions being passed on through our language, lessons, and learning?* Teachers might ask: *How are learning relationships in our classrooms influenced by root metaphors, deeply rooted cultural assumptions, and value-hierarchized dualisms?* Utilizing an ecocritical framework, ask: *What does it mean to teach toward the abolition of the dominance of problematic root metaphors, superior/inferior (either/or) thinking and decision making and in support of ecological, anti-racist, and feminist teaching?*
- Engage in critical and ethical examinations of community. As notions of community are all too often defined in terms of white supremacy, patriarchy, and human-centered exclusion, it is important to reconsider community in terms of who and what are included in our definitions of this construct and how those definitions contribute to either supporting or undermining the need for all beings to coexist in peace. For example: *Who and what might*

we be ignoring when we think about who is considered in decisions in our neighborhood community? What animals and plants live and make their homes in our community, and how do we depend on them? What are we doing to practice reciprocity with our diverse human and more-than-human neighbors? How might we work together to revitalize and strengthen the commons? Furthermore, imagine how students could be prompted to recognize, consider, and value diverse cultures and species when learning about citizenship and voice. Teachers can work with students to imagine how decisions might be made that consider more broadly that all voices matter. *Given the historical dominance of White male voices, how important is it to center diverse voices, ideas, abilities and experiences? How might we begin to expand our understanding of voice to mean more than utterances of human language systems that rely on words?* Consider engaging students in learning to listen and be responsive to diverse language systems like breathing, smell, diverse sounds, gestures, as well as weather, climate, water, soils, birds, insects, fungi, forests, and other mammals and animals. Specifically, work to identify—or seek out—a more-than-human teacher (something/ one you learn from and intentionally engage in a learning relationship). At first, this is just about making a commitment to learning from this different kind of teacher-student relationship in a way that interrupts habits and assumptions of anthropocentrism and human-supremacy. Then, journal over the course of the semester to share your experiences with how you learned from, and in many cases learned to listen to, your ecological surroundings. Commit to learning about the ways in which oppressed communities have suffered and survived the extreme violence of white-male heteronormativity without unwitting complicity with these atrocities.

- Engage in examining community in terms of inclusion and the diverse ways in which our living relationships can be recognized, respected, and represented through teaching and learning among all community members. Specifically, engage in recognizing the role activist networks play in alleviating and eliminating unjust suffering in our communities. Build networks of solidarity with these organizations. How can single-issue social justice groups make alliances with other social and/or environmental justice groups? For example, what commonalities and bridges might there be between Planned Parenthood, the BLM movement, and climate change activism? A promising entry point for this intersectional work is food and water as important aspects of the cultural and environmental commons. Everybody eats, and we need water and are made of water! How could teachers connect with and start conversations between organizations, our classrooms, and local communities? As teachers, this organizing starts with learning about who is already attempting these conversations and how their attempts might be supported. Additionally, think about shifting classroom communities in order to encourage collaboration rather than competition. As teachers, focus lessons on fostering and developing skills of community collaboration rooted in mutual aid and interdependence. School and community gardens can be a site of such teaching and learning. Ask: *How might we learn to teach from the Black Lives Matter at Schools resources together with local food justice work or water protector lessons?* Or, *What might it be like to teach or be a student in Detroit, Flint, Chiapas, Cuba, or Rojava?*

- Engage in supporting the diverse approaches to taking up resistance and healing from Western industrial culture and, in solidarity, show respect for epistemologies that differ from the current dominant discourses of Western industrial culture. Support the ways in which diverse forms of resistance work to challenge value-hierarchized dualisms that perpetuate value-hierarchized thinking. For example, explore the ways in which local groups in your community are fighting against past and present acts of colonization, both in the United States and internationally. Imagine how educators might teach lessons that challenge, and question progress rooted in consumerism and global market ideology. For example, have students explore barter and trade in efforts to make explicit that economic systems do not require capitalism to exist. Have students identify non-monetized activities that exist in the community. Additionally, have students explore the gendered dimensions of non-monetized caring work and how that work upholds and strengthens communities.

Above all, and in addition to the attempts to disrupt institutionalized Western industrial culture that is perpetuated through and within educational spaces, we can also commit to the daily effort of making critical friends with other humans and non-human animals, engaging with adversaries, and sharing stories of hope and resilience while critiquing local and international relations of domination and hierarchies. In such volatile, authoritarian times, it is important that critical educators challenge dominant perceptions of what currently constitutes schooling, education, and knowledge to collectively imagine with open hearts and minds possible alternatives. Through friendship and critical dialog, we can resist white-male heteronormative human-supremacy and reject the illusion that as humans we are separate from and superior to each other and all other beings on the planet. We challenge human supremacy when we make friends with other humans and especially when we teach one another to make friends with more-than-humans—be it animals, trees, a river, the food that we grow, or the mycelial networks in the soil that give us life. The point is that we learn compassion, dependency, and different ways of listening and communicating when we understand in an ecological sense what it means to be friends—to recognize and value that we are in relationship with a vast variety of diverse beings and that we owe our existence to these relationships. We learn what it means to belong without framing that understanding within anthropocentrism, whiteness, patriarchy, or capitalism. Rather, belonging becomes the relationality that we enact in our everyday lives, existing within healthy and mutually supportive ecosystems. It is through these friendly and mutually sustaining relationships that we learn to overcome the isolating ills of Western industrial culture, and we are called to action with diverse others to teach in support of living systems and together face what Baldwin (1963) suggested will be met with "the most fantastic, the most brutal, and most determined resistance" (p. 42). He reminds us that while teaching in these dangerous times of crisis: "The obligation of anyone who thinks of himself [sic] as responsible is to examine society and try to change it and fight it—at no matter what risk. This is the only hope society has. This is the only way societies change" (p. 42). Chet was a teacher, a friend, and without a doubt a curriculum scholar whose work is both far reaching and provocative and embodies the kind of tenacity and persistence that anyone who thinks of themselves as responsible ought to be examining society and trying with all their abilities and resources to change it and fight it. Chet Bowers, most certainly right up to his final living moments was engaged in precisely that work—may his work among its imperfections and tensions continue to find meaning and life in not only the authors and readers of this book but beyond academia and in the imminent cultural changes on the horizon for humans on this planet. One last time from our friend, the teacher who reminded us often: "The ecological crisis is a cultural crisis"—may our collaborative and creative efforts shift society and may our diverse communities find resolution to the unjust suffering we are all working to abolish. "Enough!"[1]

Note

1. Chet often ended his emails and conversations with me on these topics with the single stern word "Enough." I am guilty of often asking question after question after question and he'd often for hours indulge in person or in his long E-mails. I miss him and hearing/reading his signature word to end our conversations and selfishly I feel comforted by closing this collection of chapters.

References

Baldwin, J. (1963). A talk to teachers. *The Saturday Review*. December 21, 1963, pp. 42–44.

Bowers, C. A. (2002). Toward an eco-justice pedagogy. *Environmental Education Research, 8*(1), 21–34.

Braslow, S. (2020). Black lives matter estimates that as many as 100,000 protesters gathered in Hollywood on Sunday. Los Angeles Magazine. June 8, 2020. Retrieved from https://www.lamag.com/citythinkblog/hollywood-protest-sunday/

Center for Disease Control. (2020). CDC COVID Data Tracker. Retrieved from https://www.cdc.gov/covid-data-tracker/#cases

Cochran-Smith, M., Feiman-Nemser, S., McIntyre, D. J., & Demers, K. E. (Eds.). (2008). *Handbook of research on teacher education: Enduring questions in changing contexts.* New York, NY: Routledge.

DeLeon, A. P. (2019). *Subjectivities, identities, and education after neoliberalism: Rising from the rubble* (Vol. 41). New York, NY: Routledge.

Lupinacci, J., & Lupinacci Ward, M. (2017). (Re)imaginings of "community:" Perceptions of (dis)ability, the environment, and inclusion. In A. J. Nocella II, A. George, J.L. Schatz, & S. Taylor (Eds.), *The intersectionality of critical animal, disability, and environmental studies toward eco-ability, justice, and liberation* (pp. 63–78). New York, NY: Lexington Books.

Lupinacci, J., & Happel-Parkins, A. (2016). Ecocritical foundations: Toward social justice and sustainability. In J. Diem (Ed.), *The social and cultural foundations of education: A reader* (pp. 34–56). San Diego, CA: Cognella.

Lupinacci, J., Happel-Parkins, A., & Turner, R. (2019). Ecocritical pedagogies for teacher education. In M. Peters (Ed.), *Encyclopedia of teacher education.* New York, NY: Springer.

Lupinacci, J., Parkins-Happel, A., & Turner, R. (2018a). Ecocritical scholarship toward social justice and sustainability in teacher education. *Issues in Teacher Education, 27*(2), 3–16.

Lupinacci, J., Happel-Parkins, A., & Ward Lupinacci, M. (2018b). Ecocritical contestations with neoliberalism: Teaching to (un)learn "normalcy". *Policy Futures in Education, 16*(6), 652–668.

Martusewicz, R. A. (2019). *A pedagogy of responsibility: Wendell Berry for ecojustice education.* New York, NY: Routledge.

Martusewicz, R., Edmundson, J., & Lupinacci, J. (2020). *EcoJustice education: Toward diverse, democratic, and sustainable communities* (3rd ed.). New York, NY: Routledge.

Parker, K., Manasce Horowitz, J., & Anderson, M. (2020). Amid protests, majorities across racial and ethnic groups express support for the Black live matter movement. Pew Research Center: Social & Demographic Trends. Retrieved from https://www.pewsocialtrends.org/2020/06/12/amid-protests-majorities-across-racial-and-ethnic-groups-express-support-for-the-black-lives-matter-movement/

Stengers, I. (2015). *In catastrophic times: Resisting the coming barbarism.* London, UK: Open Humanities Press.

Turner, R. J. (2015). *Teaching for ecojustice: Curriculum and lessons for secondary and college classrooms.* New York, NY: Routledge.

Index

Note: Page numbers followed by "n" refer to notes.